The Sustainability of Operations

Past, Present, Future

Joris van de Klundert

now
the essence of knowledge

Published, sold and distributed by:
now Publishers Inc.
PO Box 1024
Hanover, MA 02339
United States
Tel. +1-781-985-4510
www.nowpublishers.com
sales@nowpublishers.com

Outside North America:
now Publishers Inc.
PO Box 179
2600 AD Delft
The Netherlands
Tel. +31-6-51115274

ISBN: 978-1-63828-284-6
E-ISBN: 978-1-63828-285-3
DOI: 10.1561/9781638282853

Suggested citation: Joris van de Klundert. (2024). *The Sustainability of Operations: Past, Present, Future.* Boston–Delft: Now Publishers

Table of Contents

Foreword to "The Sustainability of Operations: Past, Present, Future"

How we organize our production, our distribution, and our services is conveniently summarized in the word "operations". The efficiency and robustness of carrying out operations have become standard criteria by which operations can be measured. And indeed, the subject of operations has received its share of scientific interest, and corresponding books. Until now, however, only few books consider operations from a global sustainability point of view. This book does: it discusses planetary boundaries and focusses on social inclusion. And no book has started at the very basics, i.e., at the roots of how human beings started to organize "things". This book does: it even starts with discussing operations of ecosystems prior to the existence of humankind. This text clearly reveals how operations are impacting planetary boundaries and social inclusion; and this text clarifies how operations came to be as they are. Based on rigorous definitions of relevant concepts, the text takes the reader from the origin of operations to the present day. The author is not afraid to expand the traditional scope of operations (production, logistics, distribution) to include non-human operations. Various episodes in the development of operations are described: from the global cotton supply chain to the impact of the development of script – seemingly distinct developments are brought together in a coherent story. Not only have human beings shaped operations, the book shows how operations influenced human beings. In the last chapter pathways towards a sustainable future are explored.

Joris van de Klundert has unearthed a truly astounding collection of fact, implications, and insights, and assembled these into an impressive manuscript. I am sure that any reader will find something new in the book.

This book is original, thought-provoking, and a pleasure to read!

Frits Spieksma Eindhoven, October 9, 2023

Acknowledgments

Many people have contributed to this manuscript, most unintendedly and unknowingly. Especially the many students who sincerely and politely expressed their lack of understanding while I was introducing them to the 'obvious' principles of operations management have been a source of inspiration. I am indebted to them and to anyone else who allowed me to spend time on developing this book and/or listen to incoherent intermediate updates on the content while I was trying to grasp the essence of operations, operations management, and sustainability.

My gratitude also goes to Now Publishers, in particular to Alet Heezemans, Mark de Jongh, and Zac Rolnic, for their commitment and support to write this book, which is only remotely related, if at all, to Zac's original request.

I am also grateful for being generously allowed to reuse images. To Susan Allison, Institute for the Study of Ancient Cultures, University of Chicago, USA, for support with Figures 6.1 and 6.2. To Nima Nezafati and Thomas Stoellner from the Bergbau Museum, Bochum, Germany, for permission to reuse Figure 5.2. To the IPCC to grant permission for using Figures 2.4, 11.2, and 11.3. Lastly, to Paul Koch and Tony Barnoski for their support and co-creating Figure 4.4, which updates Table 1 of their 2006 publication [139].

DOI: 10.1561/9781638282853.ch1

Chapter 1

Introduction

We are operating this planet like a business in liquidation.

Al Gore

1.1 A Time for Change in the Sustainability of Operations

Over the first decades of the 21st century, humankind has increasingly recognized that it lives and works in ways that compromise the rights of future generations to meet their needs [180, 398]. Instead, the operations of humankind carve out a path toward a troublesome future with rising human inequalities, biodiversity loss, environmental pollution, and climate change. Humankind has already transgressed the boundaries of a safe, sustainable operating space and is operating in a red zone [163, 487, 543]. Further advancement in the present direction may lead to dramatic changes for planet Earth and humankind that are irreversible on timescales meaningful for society [518, 553].

Barring a handful of exceptions, the governments of the world's nearly 200 countries have ratified the Paris Agreement, which states that *"sustainable lifestyles and sustainable patterns of consumption and production... play an important role in addressing climate change"* [397]. Moreover, the United Nations has formulated and agreed on a Sustainable Development Agenda with Sustainable Development Goals (SDGs), which more widely address the sustainability of the planet and the species inhabiting it [398]. For most of these SDGs, the world is unfortunately not on track by 2020 [398]. For some goals—among which goals are in the SDG 12 domain of sustainable production and consumption patterns—humankind is not even moving in the right direction [518].

The Paris Agreement recognizes *"the need for an effective and progressive response to the urgent threat of climate change on the basis of the best available scientific knowledge"* [397]. The agreement therefore also tasks scientists in relevant disciplines to develop and provide such scientific knowledge and help guide adaptation action with urgency [163, 397, 518]. The Paris Agreement and SDG 12 explicitly point at production and therefore call on scientists in production and operations management and adjacent disciplines to contribute to finding effective responses to the urgent sustainability threats.

As will be further elaborated below, the ways of work of humankind, and in particular the production operations, form an important primary focus of sustainability efforts. This primary focus of the Paris Agreement, as well as SDG 12, additionally points at consumption and lifestyles, i.e., at the way humankind lives while not at work. Indeed, all operations of humankind, whether in production, logistics, other forms of work, at home, in between, or otherwise, are insufficiently sustainable and need urgent change. This implies a much broader scope for operations than the traditional positioning in the business realm, predominantly in manufacturing and logistics. The broad scope encompasses all operations of humankind as required to address the sustainability of humankind and planet Earth in three interrelated core domains: economic development, social inclusion, and environmental protection [398, 518].

In view of the magnitude and urgency of the challenges presented, the Sustainable Development Agenda calls for a revolution in operations [171, 178]. The World Economic Forum posits that we stand on the brink of a revolution in operations indeed. A revolution *"that will fundamentally alter the way we live, work, and relate to one another"* [517]. These developments and qualifications have, however, not primarily emerged within the sustainability realm, yet regard the fourth industrial revolution (4IR). The revolutionary nature of the 4IR lies in the expectation that it will be *"unlike anything humankind has experienced before"* in terms of *"scale, scope, and complexity"* [517]. It brings a blend of physical, digital, and biological changes to the way we work and live. These changes extend well beyond the manufacturing and supply chain domains of Industry 4.0, which is starting to receive attention in conjunction with sustainability [307].

The urgent demand for a revolution in operations to undo the unsustainable developments in the operations of humankind thus coincides with a nascent industrial revolution, the 4IR [487, 533]. Together, this demand pull and technology push provide a unique, rapidly closing time window of challenge and opportunity for the present generation of humankind to transition toward sustainable operations. "*We are the first generation to feel the impact of climate change and the last generation that can do something about it*" [415]. There can hardly be a more compelling reason to prioritize attention to operations. An analysis of the present operations is needed to understand the unsustainable nature of existing operations and how they have come into existence. Next, the solution to the existing challenges requires to develop essentially different ways of working and living.

1.2 Research Aims

As evidenced by the numeral four in 4IR, humankind has revolutionized operations before. In fact, several important threshold developments in operations predate humankind and have shaped humankind. Once upon a time, predecessors of humankind living in trees transitioned to living on the ground. This transition brought revolutionary changes in operations such as walking upright and—in a later stage—stone tool manufacturing and the use of fire [13]. The many generations of humankind that worked and lived after this unnumbered and unnamed "revolution" are known by their operations: hunter-gatherers.

The transition from hunting and gathering to agricultural operations that occurred more than 10,000 years ago is known as the agricultural revolution. As covered extensively in Chapter 5, this revolution dramatically changed how humankind worked and lived as it initiated the first civilizations and thus had unprecedented impact on economic development, social inclusion, and environmental protection in ways that continue until today [634]. For instance, the agricultural sector was responsible for more than one-fifth of global greenhouse gas (GHG) transmissions in 2018 (when including forestry and other land uses), as highlighted in Table 1.1 [198, 327, 553].

The first industrial revolution of the late 18th century introduced the use of the fossil fuel coal to generate steam as a power source for machine operations. The second industrial revolution of the 19th century added fossil fuel oil, the combustion engine, and electricity as energy sources to power the machines used in the operations of work and life. The GHGs emitted by the fossil fuel-powered energy sector and industry are causally linked to the transgression of various planetary and social boundaries [543]. Table 1.1 shows that industry and the energy sector together are the sources of 58.7 percent of all GHG emissions and of 45.8 percent when reallocating emissions of energy supplied by the energy sector to their users [327]. Chapters 7 and 8 cover these two industrial revolutions and their impact on sustainability.

Table 1.1. 2018 relative greenhouse gas emissions for economic sectors and the operations of life [327, 480, 553].

Sector	Direct emissions	Indirect included
Agriculture, Forestry, Other Land Use	22.1	22.3
Energy	34.2	11.0
Industry	24.5	34.8
Services	8.5	12.8
Operations of Life	11.5	19.0

The second industrial revolution also brought the combustion engine and fossil fuel transportation devices as used in the transportation services sector and in the operations of life. Combining the GHG emissions of office buildings and transportation, 12.8 percent of all GHGs are emitted by the service sector, which has grown so rapidly in the 20th century and is covered in Chapter 9. The remaining 19 percent are emitted by the operations of life, such as heating or cooling homes, cooking, traveling, and communicating.

The upper limit for a planetary GHG boundary in a safe operating space is set at 450 atmospheric CO_2 parts per million (ppm) [322, 456]. Starting from a pre-industrial level of 280 ppm, the more than thousand-fold emission growth since has caused atmospheric CO_2 levels to surpass the 420 ppm level in 2023 [321]. At current rates, human operations are likely to cause a transgression of the boundary between 2030 and 2035, causing rises in the average global temperature above which irreversible impacts on the ecosystems of planet Earth are highly likely to occur, which will severely impact society as well [456, 522, 543].

In addition to the GHG boundary, there are several other planetary and social boundaries of a safe operating space that current human operations have transgressed or might transgress in the coming decades. These boundaries, for instance, concern the abundant, unsustainable use of the biochemicals nitrogen and phosphor in agriculture, the production of harmful novel entities such as microplastics, and the loss of habitat from deforestation, which causes biodiversity loss. Bearing in mind that these boundary transgressions are directly caused by human operations, our research aims are:

Research Aims:

1. To analyze the history of the operations of humankind and its impact on sustainability. This analysis can provide insight into the logic and value of existing operating models and the enablers and barriers of the transformation in operations that is required to return to a safe and just operating space.

2. To take inventory of the present urgent sustainability challenges as caused by operations and of the ongoing technological innovations of the 4IR, and to establish the relationship between them.
3. To explore how to redesign and manage human operations to form sustainable future ways of working and living for humankind, in particular by leveraging the technologies of the 4IR.

The prime focus will be on sustainable operations from the perspective of humankind and planet Earth. Operations management, which typically adopts an organizational perspective, will be a secondary focus. Thus, while not primarily based on operations management, this study explicitly relates to existing operations management practices and literature. From the organizational perspective of the operations management literature, the three core sustainability domains of economic development, social inclusion, and environmental protection have been defined as the triple bottom line [178, 591]. From Chapter 3 onward, the final section of each chapter relates to operations management perspectives. In line with the research aims, these perspectives mostly consider how operations can help mitigate sustainability challenges. The operations management challenges associated with adapting to changes caused by unsustainable operations, such as the effects of drought on agriculture or the effects of extreme weather events on global supply chains, are beyond the scope of this research [163].

To properly position our unusually broad perspective of operations and adopt a corresponding multifaceted sustainability perspective, Chapter 2 introduces and examines the three subjects of operations, operations management, and sustainability. Chapters 3 to 9 follow the historical timeline of operations until present in pursuit of the first research aim. The reader may find this broad historical overview of the development of human operations of interest in itself and appreciate the extensive connection to sustainability that is highlighted in a separate section in each chapter. Chapter 10 addressed the second research aim, covering the current sustainability challenges, the 4IR, and their interconnection. Chapter 11 is rooted in current evidence and science, yet with the aim of identifying new ways of working and living for humankind and future sustainable operations that purposefully leverage the 4IR.

DOI: 10.1561/9781638282853.ch2

Chapter 2

Sustainability and Operations

Problems cannot be solved at the same level of awareness that created them.

Albert Einstein

2.1 Definition and Scope of Operations

The construct *Operation* has multiple, distinct meanings. It came into use in the late 14th century and primarily refers to the *performance of a practical work* and to *the action of functioning* [380, 428]. The word has Indo-European roots and reached present-day English via the latin noun *opus*, which means "*work, effort, product of labor, work of art,*" the plural "*opera (also activity, effort)*" and the verb "*operari (to work, be efficacious, produce)*" with the participle "*operatus (busy, engaged, occupied)*" [380]. The word opus has remained in use to refer to works of literature and musical compositions.

The word operation(s) has remained commonly used for multiple centuries, for instance, when referring to surgical or military operations. The word operations is plural, indicating it is intuitively and implicitly used to refer to multiple actions or performances. The singular *operation* has also remained in use to refer to a single activity or performance, for instance, when referring to a small business or part of a business or to an arithmetic operation in mathematics and computer science.

The example of arithmetic operations reminds us that computers perform operations and even have operating systems. Hence, the set of actors executing operations is not restricted to humankind. While computers have been introduced during the recent third industrial revolution to perform operations, the deployment of water, steam, coal, oil, gas, and electricity-powered machines of the preceding two industrial revolutions are additional examples of nonhuman operations. Going back further in time, we may note that domestication of animals to assist with operations goes back at least to the agricultural revolution. Animals have engaged in operations to provide physical strength, horse powers, and have produced food such as milk and honey. Animals too have operations, and Chapter 3 analyzes animal operations and their sustainability.

One may continue this thread further back toward a possible origin of operations. Before the agricultural revolution, hunter-gatherers already gathered products of animal labor, such as honey produced by bees. If operations are not exclusive to humankind, then when did operations start? What else can perform operations? Would plants, which can florish, grow, and capture carbon, have operations? What about bacteria, in view of their ability to cause illness, contribute to digestion in our intestines, or degrade plastic [520]?

We will not consider tracing the path toward the origins of operations beyond the emergence of life on planet Earth. This excludes, for instance, the operations of the Big Bang, if any, or operations implied by religious views on the origin of planet Earth and humankind [216, 597]. The presented analysis will be based on views that have broad scientific support, are inclusive, and are practically relevant for the sustainability challenges ahead.

In an early definition, Churchman regarded operations research as *the securing of improvement in social systems by means of scientific methods* [114]. This definition partially matches the call for science in the Paris agreement, avant la lettre. The definition answers the call to address sustainability challenges regarding economic growth and social inclusion, yet disregards environmental protection.

Interestingly, the definition only implicitly addresses operations through its suggestion that operations form a mechanism toward improvement in social systems. This view is continued in later definitions, some of which refer to "*the operations of a system*" without further defining operations [380, 467].

Sustainable operations have recently started to receive explicit recognition within the operations research realm [286]. In addition, the emerging scientific discipline of sustainable operations management has started to "*map the territory for sustainable OM*" [303] and pursue "*social, economic and environmental objectives the triple bottom line [TBL] within operations of a specific firm and operational linkages that extend beyond the firm to include the supply chain and communities*" [591]. Especially in the domains of manufacturing operations and supply chains, sustainable operations management and operations research are increasingly receiving attention [225, 303, 591], as is also the case for the 4IR and Industry 4.0 [307].

Our research aims necessitate a broadly scoped definition of operations that encompasses sustainability in the domains of economic development, social inclusion, and environmental protection. As has become clear from Table 1.1, the current sustainability challenges stretch beyond the boundaries of organizations, beyond the boundaries of supply chains, and even beyond the boundaries of social systems. Many of the most severe challenges relate to the ways we live and interact with the environment. To address such challenges, operations need to be considered within ecosystems, which form the context in which social systems function and interact with.

2.2 Ecosystems

2.2.1 Definition of Sustainable Ecosystems

With a view toward the 4IR and building on commonly adopted previous definitions, we define after [152, 462]:

Definition: An **ecosystem** consists of

1. living organisms (including humankind, animal species, plants, bacteria, et cetera),
2. physical processes,
3. products, i.e., non-living elements created by physical processes conducted and/or controlled by living organisms, and
4. a context of other non-living elements in which the living organisms and products reside and the physical processes occur.

The definition above differs from previous ecosystem definitions because it distinguishes products created by living organisms from other, typically pre-existing, non-living elements. This distinction will be instrumental in examining the timeline of operations and the sustainability of operations, in particular in relation to the emerging 4th industrial revolution.

The definition of products to refer to other non-living elements differs slightly from common use. For example, seeds, plants, and flowers are not considered products in this definition. A branch of a tree is a part of a living organism when on the tree, is a non-living element of the context when it falls off the tree after a storm, and is a product when picked up from the ground by a chimpanzee that uses it to "fish" termites after stripping off the leafs (see 3.4). Fossil fuels are elements that are part of the context until they are mined and processed under the control of humans, which converts them into products.

Products can be created by living organisms for a direct reward, for example, when cooking food to nourish a family. Other products, however, are produced for an indirect reward, for instance, when being used as resources in a physical process, as is the case for ovens or casseroles.

The use of wooden sticks after stripping the leaves by chimpanzees when hunting for termites is an example of the indirect reward use of products as tools. We define **tools** as products or other non-living elements used in support of conducting a physical process. One may notice that this definition allows for recursion, as tools can be used to produce tools. Chapters 4 onward show that each revolutionary development along the timeline of operations came with its own defining tools. Stone tool manufacturing has been a first threshold development for the predecessors of humankind. The agricultural revolution brought the plough, and the light bulb has become an icon of the second industrial revolution.

Tools can support physical processes for which humans or other living species provide energy in the form of muscular power, as well as physical processes conducted by machines. We define **machines** as products created to conduct physical operations using nonhuman sources of energy, possibly in addition to human energy. We assume machines (co-)conduct physical processes and operate, directly or indirectly, under the control of living organisms, typically humans. This control is, for instance, indirect when a machine is controlled by another machine (e.g., a computer), which is under human control. Machines can produce consumption goods as well as indirect reward products such as tools or even machines.

Thus, in the remainder, a broom is a tool and an electrically powered vacuum cleaner is a machine, regardless of whether it is operated by a human or is a robot vacuum cleaner. Water mills, sailboats, pizza ovens, steam locomotives, and nuclear reactors are also machines. Tools and machines together define **technologies**. These examples show that the machines developed before the industrial revolution typically used renewable energy sources such as wind. However, the nonsustainable use of wood for combustion and ship construction already occurred well before the industrial revolution, as discussed in Chapters 4,7. The industrial revolution essentially introduced numerous fossil fuel and electricity-powered machines into the planetary ecosystems for the purpose of conducting physical processes and operations under human control.

It is worth noticing that these definitions of tools, machines, and technologies are helpful for the purpose of examining the sustainability of operations but differ from commonly encountered historical definitions of machines. More than 2,000 years ago, for example, Archimedes considered wedges and screws to be basic machines [112]. The other four of the six historically identified basic machines are the lever, the pully, the inclined plane, and the wheel and axle [111]. In our definition, these are tools that can be used by living organisms and by machines.

The practical definition of technologies provided above is narrow as it is limited to tangible ecosystem elements.

Ecosystems can be defined and scoped at various levels. In the remainder, we often consider planet Earth as a single ecosystem. On occasion, we distinguish relevant smaller (sub)ecosystems, ranging from oceans and continents to Mesopotamia, the Amazon rain forest, the waters surrounding the island of Manhattan, or an African e-waste processing site. The following definition implicitly encompasses the social and economic sustainability perspectives of the ecosystem:

Definition: An ecosystem is **a sustainable ecosystem** if the needs of the present and future generations of living organisms within an ecosystem, including the economic and social needs of humans living in the ecosystem, are not compromised by the present physical processes.

2.2.2 An Ecosystems-based Definition of Operations

With the definition of sustainable ecosystems at hand, we now define operations and sustainable operations.

Definition: Operations are the physical processes in an ecosystem as controlled and possibly conducted by the community of living species in the ecosystem.

Depending on the boundaries of the ecosystem considered, this definition may exclude some physical processes external to the ecosystem. The definition is consistent with the aforementioned disregard for any operations occurring before the emergence of life on planet Earth.

We now specifically define:

Definition: Human operations within ecosystems are the physical processes conducted and/or controlled by the community of humans in the ecosystem.

Thus, in addition to operations directly conducted by humans, operations such as a donkey pulling a cart, a ship sailing the ocean, a steam locomotive pulling a train, and a nuclear reactor producing energy are also considered human operations.

We now generally define:

Definition: Operations within an ecosystem are **sustainable operations** if they do not compromise the needs of the present and future generations of living organisms in the ecosystem, including the economic and social needs of humans in the ecosystem.

Operations that are not sustainable, as per the definition above, will be called **unsustainable**. This definition of sustainable operations differs essentially from definitions within research disciplines such as operations research and definitions within the business management realm [286, 303, 591].

In pursuit of our research aims, human operations will receive the most attention in the remainder of the analysis. It will therefore be valuable to separate human operations from those conducted by other living organisms in ecosystems beyond human control. A human-centered view of these operations conducted by nonhuman organisms defines them as a set of services that "*sustain and fulfill human life*" [138]. Building on recent literature and our previous definitions, we would rather define [10]:

Definition: Ecosystem services are the physical processes within ecosystems conducted by nonhuman organisms in the ecosystem without being controlled by humans.

These ecosystem services encompass an almost infinite variety of operations, several of which will receive attention in subsequent chapters because of their interactions with human operations. For instance, they include services conducted by wild birds and insects, such as the dispersion of seeds and pollination. They also include operations by naturally occurring microorganisms and bacteria, such as soil fertilization and the detoxification and decomposition of waste [139]. We may further think of the photosynthesis process through which plants process carbon dioxide and water to produce oxygen. Of all operations taking place on planet Earth, human operations are only a small subset.

Chapter 10 extensively discusses the ecosystem services presently threatened by transgressions of the boundaries of a safe operating space for the planet and humankind [151, 487]. These ecosystem services are threatened by the effects of the technologies used in human operations since the industrial revolution. Chapter 11 explores how the adoption of existing and future technologies of the 4th industrial revolution can reverse these effects and help mitigate the threats human operations pose to ecosystem services.

2.2.3 Ecosystems Dynamics and Ecosystem Engineering

Ecosystems are dynamic. External events, such as meteorite impacts or volcanic activity, can initiate physical processes that profoundly disturb ecosystems. Under such circumstances, evolution naturally favors species of living organisms that are responsive to such changes. For this purpose, individuals of these species need to continue to operate effectively, foremost by keeping themselves alive long enough to produce and raise offspring.

Organisms and species can promote their survival probabilities by conducting operations that modify the ecosystems in which they live to become more favorable for their longevity and reproduction. Such operations have been referred to as ecosystem engineering and as niche construction. More formally, **ecosystem engineering** refers to the *physical modification, maintenance, and creation of habitats by living species* [289]. **Niche construction** has been compactly defined as *the modification of selective environments by living organisms* [325, 416]. We will use the two interchangeably.

It is worth noting that ecosystem engineering (and niche construction) are ecosystem dynamics caused by internal processes rather than by external events. Obviously, human ecosystem engineering has been a very significant source of ecosystem dynamics, especially in the last few centuries, to the extent that the sustainability of many planetary ecosystems is at risk or might be soon.

The ecosystem engineering operations of humankind and other species often involve the amelioration of safe access to food for members of a species. The food chain within an ecosystem can therefore provide a helpful perspective to understand ecosystems. Food chains, in turn, can be understood as systems that provide energy and calories, which will appear as an important theme throughout the timeline of operations (see also [104]).

Plants can be viewed as the basis of food chains. Plants consume energy, e.g., for plant growth, using environmental resources such as water, oxygen, light, carbon, nitrogen, phosphor, and warmth. With their physical processes, their operations, plants can capture carbon as a resource and also produce energy, which they may store in their roots, branches, leaves, fruits, et cetera.

Following the food chain paradigm, the energy produced by plants provide the nutrition for herbivores [104]. However, not all of the energy produced by plants is consumed by herbivores. Much of the net energy produced by plants—i.e., the surplus of the energy produced over the energy consumed by plants—enters the soil as organic matter. In cases of wild fires and human-made fires, the carbon produced by plants forms the energy source for combustion, and the carbon is emitted into the atmosphere.

Keeping the food chain perspective on the energy flow in ecosystems, herbivores are the second stage. Herbivores consume plants and use the energy thus consumed for their own operations, in particular for reproduction and subsistence. Dead or alive, herbivores subsequently form a source of energy for other species, such as predators and bacteria. Predators, in turn, consume energy for reproduction and subsistence and may form a source of energy for other predators or bacteria, et cetera.

Together, the net energy productions of the species involved in the food chains of an ecosystem define the **net carbon production of an ecosystem** [104]. Remains of plants and animals not consumed in the food chain typically end up as biomass in the soils or on the seabeds of the ecosystems they have lived in. Over millions of years, such biomass forms thicker and thicker layers undergoing biological processes (for instance, by bacteria) and subsequent geothermal processes involving pressure and heat. As a result of these physical processes, the carbon originally contained in the organic sediments forms a high caloric component of resulting material resources such as coal (largely from land-based biomasses), petroleum (largely from marine biomasses), and related gasses [83]. Because of their organic origin, these energy resources are commonly referred to as **fossil fuels**.

Thus, the physical processes of an ecosystem can even transform the net carbon produced and contained in the remains of the community of living organisms (the first component) into non-living elements in the environment of the ecosystem (the fourth component). These processes, however, happen at timescales that are much larger than the timescales that are meaningful for the organisms living in ecosystems.

This brief energy-oriented synthesis of food chains in ecosystems provides a first picture of the energy systems in ecosystems. However, it does not yet include the use of energy by humans in human operations. As subsequent chapters reveal, the operations of humankind have increasingly adopted the use of living and non-living elements of ecosystems as energy resources to create products (and services) while constructing niches more conducive to human life. The net carbon emissions of operations to produce a specific product (or a service) are called the **carbon footprint** of the product (or service).

Carbon footprints can also be defined for living elements of ecosystems, and in particular for humans and human populations, as the net carbon emissions of a set of allocated operations. We shall see in subsequent chapters how the carbon footprint of the human population increased as its niche construction practices progressed and humankind secured a position at the top of the food chain. The carbon footprint of humankind has become unsustainable after transitioning to operations that use fossil fuel-powered technologies. Further advancement in this direction may lead to dramatic changes for the ecosystems of planet Earth and humankind that are irreversible on timescales meaningful for society [518, 553].

2.3 Measures for the Sustainability of Operations

The sustainability challenges of the operations of humankind are expressed in terms of the impact on economic growth for the poor, on social inclusion, and on the environment. To analyze these impacts of operations, we review and elaborate on these three domains in relation to operations, building on current scientific frameworks and on the measures developed in support of the SDGs for this purpose [398, 518].

2.3.1 Economic Development

Gross domestic product (GDP) is a widely used indicator for economic performance that intends to directly express the sum of the gross values of all goods and services produced by the population within an economy [37, 571]. Economic development can then subsequently be defined in terms of economic growth: the change in real GDP, i.e., the change in GDP at constant prices [37]. The change in real value produced per population member at a constant price can then be expressed via the real GDP per capita [37, 571].

Economic growth for the poor is an explicit SDG and promoted in the Paris Agreement [37, 397, 398]. SDG 8 entails an annual 7 percent increase in real GDP per capita for less developed countries [398]. As this reflects an increase in the value of goods and services produced, this SDG presents a direct objective for the per capita operations conducted within these countries. A closely related measure considered among the SDGs is labor productivity, i.e., the GDP per (full-time) worker.

Figure 2.1 shows estimates of global GDP (in millions of 1990 USD) and global GDP per capita (in 1990 USD) for the last 2,000 years. Applying a logarithmic scale, the super linear trend evidences exponential economic growth and recent growth acceleration, both per capita and globally. On average, the output of operations per human is 15 times more valuable than it was 2,000 years ago (assuming that estimates for year 1 are accurate). From the perspective of planet Earth, this per capita GDP growth is amplified by a factor of 30 because of concurrent exponential population growth (assuming estimates for year

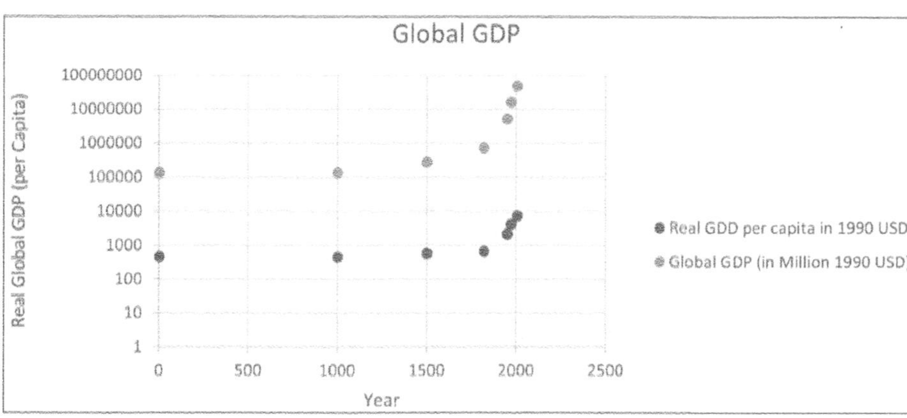

Figure 2.1. Real global GDP (per capita), source [85, 357].

1 are accurate). Planet Earth has accommodated and absorbed 2,000 years of exponential growth in human operations, which has resulted in the real value produced growing by a factor of 450. What is more, according to human measures, this growth is considered insufficient as the SDGs call for further exponential growth in some countries. Chapter 10 discusses the present and future sustainability of these developments.

While being widely accepted and utilized, GDP-based measures for economic growth—and more generally for economic development—have important shortcomings that are relevant for our analysis [571].

First, there are various critiques of the accuracy of reported GDPs, further exacerbated by the neglect of informal production [571]. Informal production includes both operations in the form of unregistered activities considered "labor" (the way we work) as well as other human operations (the way we live). A tomato bought in the supermarket is included in the GDP. A tomato bought informally from the nearby tomato farm is not. A home-grown tomato from the kitchen gardens is not either.

Likewise, it is important to be aware that GDP is a measure of production by a certain population over a certain period and may differ from the income for the corresponding population over the same time period. It is the income, however, rather than the production, that matters for welfare and well-being, provided that it can be converted into goods and services that promote welfare and well-being. The relationship between GDP growth and increases in welfare and well-being is ambiguous. It is easily seen that equal relative increments in personal income will not forever translate to equal relative increases in well-being. Recent scientific evidence on (the lack of) correlation between the GDP and the well-being of populations is presented in Chapter 10.

The prime reason for GDP and income to be included in the SDGs is to reduce poverty and "*leave no one behind*," which forms "*the central, transformative promise of the 2030 Agenda for Sustainable Development*" [398, 503]. The commonly reported real (per capita) GDP, however, is a population average, which can be quite different from incomes obtained across a variety of population members. Even if the average real per capita GDP increases, the real income of some members of a population may decline. Economic growth can thus be unsustainable if the population members with the lowest incomes are left behind. Adverse relationships between GDP growth and economic growth for the poor can occur locally, nationally, and between nations and continents. In agreement with the SDGs and the Paris Agreement, the presented analysis will therefore often focus on (sub)populations that might be left behind, and particularly on developing countries, as is necessary for poverty relief. This focus aligns with the sustainable development objective of social inclusion further discussed below [397, 398].

In relation to the third sustainable development objective of environmental protection, it has been observed that the measure of GDP and many other productivity measures disregard exhaustion of natural resource reserves and other impacts on the environment [571]. While the environmental perspective of such ecosystem effects is addressed separately by sustainability indicators constructed to this purpose (see subsection 2.3.3), this still leaves the economic value of natural resources (as a form of capital) and costs of restoring ecological damage unaccounted for.

After reflecting on past developments and with a deeper understanding of the present sustainability challenges and the present context, Chapter 10 revisits measures for economic development that aim to have more validity for today's sustainability challenges than GDP. Along the timeline of operations in Chapters 3 to 9, GDP and per capita GDP, with a special eye toward the poor, will play important roles as measures of economic sustainability.

2.3.2 Social Inclusion

There is no standard measure of social inclusion that is as widely accepted as the measure of GDP for economic growth. One explanation for this absence lies in the view that social inclusion is best defined differently in different contexts [524]. On a global level, social inclusion directly relates to SDGs: 1, no poverty, and 10, reduced inequalities. Figure 2.2 shows the remarkable progress made by humankind over the last 40 years, in which the percentage of the global population living at the equivalent of 1.90 real 1990 USD per day or less decreased from over 40 percent to less than 10 percent in relative terms and from around 2 billion people to well below a billion in absolute terms.

Social inclusion is not limited to income inequalities. SDGs addressing unemployment rates, access to education and health, poverty, and income inequality also specifically target social inclusion [500]. The European Union has formulated

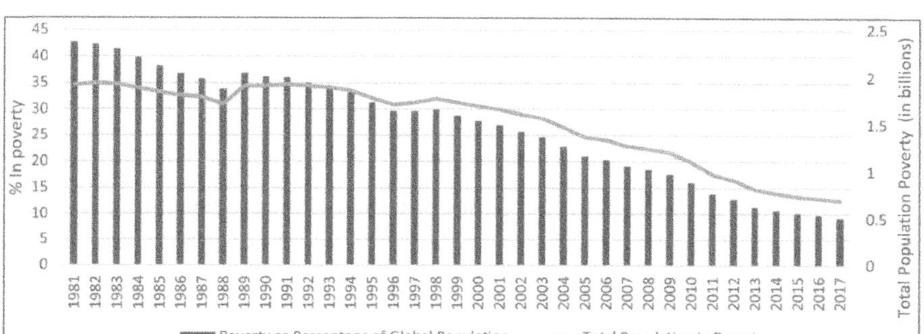

Figure 2.2. Percentage of global population living from less than 1.90 USD per day (1990 USD), source [38].

measures on these dimensions as well [27]. In fact, social inclusion and exclusion can be related to each of the SDGs.

Frameworks for a safe and just space for human development that are closely related to the SDGs in general—and specifically to environmental protection as covered below—have received considerable attention recently [412, 472, 473]. These frameworks distinguish a set of needs that apply to present and future human generations and specify social boundaries for justly and safely meeting these needs [473]. The most basic of these needs are physiological and can be related to the lower levels of Maslow's hierarchy of needs [365, 412]. Other, higher-level needs, such as education, happiness, and internet access, are also included.

Equity and equality are often considered separate social priorities among the human needs [472, 473]. Alternatively, they can be considered to apply to all human needs and this is the approach we adopt for the remainder. Hence, we define [74, 472, 473, 601]:

Definition: Social inclusion refers to intragenerational and intergenerational equity in meeting the needs of present and future human generations in terms of

- food,
- water,
- sanitation
- health,
- income,
- education,
- housing,
- energy,
- work,
- well-being,
- social networks,
- political voice,
- peace and justice,

where equity refers to the absence of unjust or unfair inequalities, as far as can be avoided, among subpopulations and generations.

Notice how this definition implies gender inequality and social equity to apply to each of the enlisted needs instead of considering them as separate needs, as is done in some other frameworks [472, 473]. The presented definition explicitly defines equity to apply to all social inclusion dimensions and is therefore aligned with preceding frameworks that express measures per dimension, such as the percentage of the population with an income above the extreme poverty line or the number of children aged 12–15 years out of school [473].

Without suggesting a ranking order, it can be noticed that the social inclusion dimensions occurring "higher" up appear more easily and objectively measurable. As we proceed "downward," the dimensions are of a more social and intangible nature, and their assessment may require individual (subjective), generational, and cultural (intersubjective) valuations.

Work is obviously a dimension that is very closely related to operations. However, our broad perspective on human operations includes the way we work and the way we live and encompasses all social inclusion dimensions in relation to the sustainability of operations. Subsequent chapters therefore cover one or more of the above needs as relevant, and Chapters 10 and 11 cover the social inclusiveness of present and future operations, especially in relation to the 4IR.

2.3.3 Environmental Protection

The basic set of environmental statistics of the United Nations distinguishes six categories of measures [157]:

1. Environmental Conditions and Quality (among which measures such as temperature, sea level, biodiversity (in numbers of species), and air quality)
2. Environmental Resources and Their Use (among which measures of mineral and energy reserves and forestation)
3. Residuals (among which GHG emissions, water pollution, hazardous wastes, and pesticides)
4. Extreme Events and Disasters
5. Human Settlements and Environmental Health (among which health consequences of indicators measured in the preceding categories)
6. Environmental Protection Management and Engagement

The set contains a total of 458 indicators, demonstrating the wide variety of impacts human operations have on the environment. In the remainder, we cannot cover these 458 indicators comprehensively or systematically. We would rather aim to cover measures when and as relevant for sustainability along the timeline of operations. To this purpose, we recall from Chapter 1 that humankind has already transgressed the boundaries of a safe operating space. The current scientific understanding of a safe operating space for planet Earth has been expressed through nine boundaries [322, 487, 543]. Below, we present these **nine boundaries of a safe operating space for planet Earth**, together with a selection of relevant measures for clarification (between parentheses) [443, 543]:

- Climate change (atmospheric CO_2 concentration in parts per million),
- Biosphere integrity (extinction rate),
- Land-system change (area of forested land as percentage of original forest cover),

- Freshwater use (global consumptive blue water use per year, locally as a percentage of river flow per month),
- Biochemical flows (various global and local P (phosphor) and N (nitrogen) flows,
- Ocean acidification (carbonate ion concentration),
- Atmospheric aerosol loading (aerosol optical depth),
- Stratospheric ozon depletion (stratospheric O_3 concentration),
- Novel entities (e.g., concentration relative to no effect concentration).

Novel entities are new substances, new forms of existing substances, and modified life forms that have the potential for unwanted geophysical and/or biological effects, such as chemicals and other new types of engineered materials or organisms not previously known to the Earth system, as well as naturally occurring elements (for example, heavy metals), resulting from human operations [543]. Examples are pesticides, CFCs (chlorofluorocarbons), microplastics, and nanomaterials [443].

The measures of the boundaries are included in the basic set of environmental statistics of the United Nations and are closely related to other measures in this set. For instance, the climate change measure of atmospheric CO_2 concentration is closely related to the measure of CO_2 emissions included in category 3, Residuals, as GHG emissions. This measure refers to the aforementioned net carbon footprint of humankind. Figure 2.3 visualizes this footprint on a logarithmic scale from 1750 onward. It demonstrates the exponential growth of CO_2 emissions since the onset of the industrial revolution.

Figure 2.4 places the CO_2 emissions within the larger category of GHG emissions over the past three decades. Figure 2.4 also shows how the GHG emissions resulting from human operations have grown by approximately 50 percent since 1990, despite the containment efforts made. GHG emissions resulting from human operations are referred to as **net anthropogenic emissions**, considering

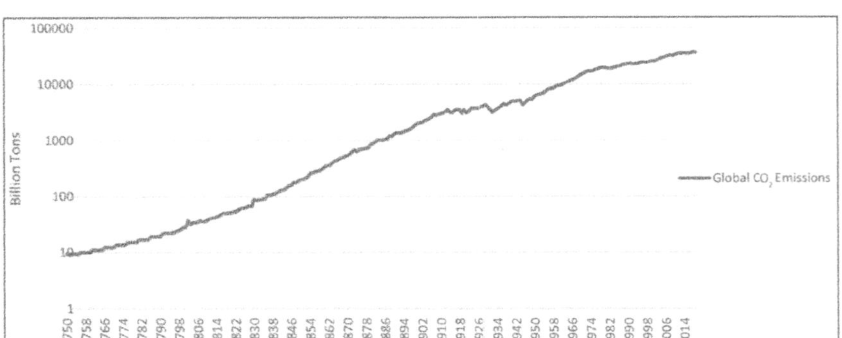

Figure 2.3. Global CO_2 emissions before the industrial revolution (logarithmic scale), source [514].

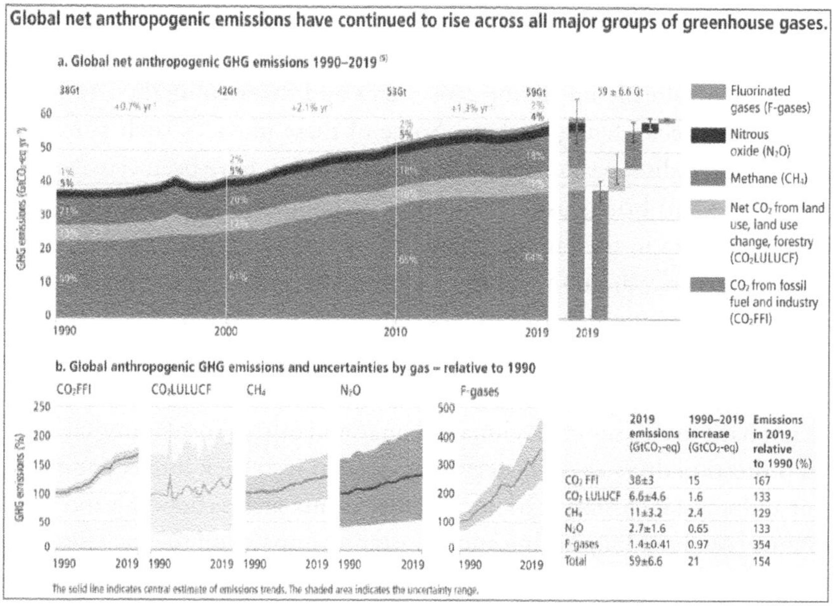

Figure 2.4. Global GHG emissions per category of gases from 1990 onward (reprinted from Figure SPM 1 in [523]).

that human operations can both increase and decrease the CO_2 emissions of the global ecosystem.

Carbon capturing refers to extracting CO_2 from the atmosphere. Carbon-capturing machines are already in operation in industry. Forests, and particularly the photosynthetic operations of the trees and plants in forests, presently capture much more carbon than these machines. Carbon capturing is relevant as atmospheric CO_2 concentrations had already grown to a level of around 400 ppm by 2014 and continued to grow afterward [480, 543]. This value moves toward the higher end of the uncertainty interval of the climate change boundary of a safe operating space, which is estimated to range from 350 to 450 ppm. Above the level of 450, human operations have transgressed the first boundary of a safe operating space, and carbon capturing is then needed to return to within the boundaries.

Carbon capturing can also serve to avoid transgressing the boundary (to avoid overshoot), and the area of forested land thus appears as a sustainability measure in the third category, Environmental Resources and Their Use, in the UN framework and is also covered by the third planetary boundary. Land use change and deforestation are forms of human operations that date back to long before the agricultural revolution, as we will see in Chapter 3.

The climate change boundary is especially important because crossing this boundary is highly likely to have a pervasive impact on other boundaries, and additional irreversible impacts on ecosystems are highly likely to occur [456, 543].

These impacts vary from extreme weather events, such as extreme precipitation events, drought, and wildfires, to sea-level change [523]. Increases in air and sea temperatures are already impacting ecosystems and threatening their suitability as habitat for the species living in them. Some of these impacts, such as extinctions of species or their disappearance from local ecosystems, have been classified as irreversible [456]. This brings us to the impact on another important environmental protection measure in the same category, Biosphere Integrity. Biosphere integrity relates to the genetic variation within species and between species. The latter is also known as biodiversity and is for instance, measured by extinction rates (see above). Of the 138,374 species of plants and animals recently assessed by the International Union for the Conservation of Nature, 38,542 were found to be threatened [124].

It is not hard to envision that climate change and other shortcomings in environmental protection drive changes in ecosystems that feed back into economic development and social inclusion. Extreme weather events such as flooding and wildfires, for instance, can have long lasting and devastating impacts on the societies affected. Drought and temperature rises already negatively impact agricultural operations, e.g., because of desertification [384]. Rises in seawater temperature reduce the effectiveness of fisheries operations. Reduced health and well-being resulting from climate change have also been reported to negatively impact economic growth [456].

The negative societal impacts of operations that fail to protect the environment vary across geographical regions and subpopulations. Approximately 3.3 to 3.6 billion people live in ecosystems that are highly vulnerable to climate change [456]. Moreover, poverty, marginalization, and other forms of inequity increase vulnerability to climate change. Hence, unsustainable development in the environmental protection domain may exacerbate existing inequities in the social inclusion domain. Extant interrelationships between environmental and social sustainability are further elaborated in Chapter 10 on the basis of the "*Doughnut model*" [331, 412, 473].

The above shows that measurement of sustainability is complex and that the many interacting measures should be considered jointly rather than in isolation. The 17 SDGs formulated by the United Nations form an integrated framework that recognizes this interdependence. For a more general discussion of sustainable development indicators, we refer to United Nations reports and recent scientific literature [157, 250, 398].

2.4 Operations Management Perspectives

The scope and definition of operations introduced above differ from commonly encountered definitions in the operations management and operations research realms that tend to address operations more implicitly. These disciplines tend to define operations within organizational contexts rather than in ecosystems. For example, operations management is commonly defined as being concerned with the design, control, and improvement of the value-creating processes of organizations that transform inputs into outputs in the form of products and services [260, 312].

Table 1.1 shows that the far majority of GHG are emitted by organizations transforming inputs into outputs in the form of products and services. GDP can be viewed as the value of these organizations, showing the importance of operations for the sustainability dimension of economic development. The operations and operating models of organizations also importantly impact social inclusion, as we shall see in Chapters 4 to 9. The operations of organizations are thus a key driver of present sustainability challenges, and operations management is therefore necessarily at the core of any solutions to these challenges. Thus, the organizational view of operations and operations management is crucial.

An empirical, practical perspective on operations management that closely matches this organizational embedding may therefore simply define it as the collection of activities conducted by operations managers (or by people in similar positions, such as operations director or chief operating officer), in function of their responsibilities. This perspective grounds operations management in actual organizational practices and follows the organizational hierarchy and structure. Operations itself can then correspondingly be defined as the activities conducted under the responsibility of the operations manager (or equivalent), such as the activities conducted in the department of operations (or equivalent).

One might argue that such an empirical grounding yields a definition of the highest practical relevance. The scientific operations management discipline then studies the corresponding empirical operations management practices. From an epistemological perspective, such a definition differs essentially from an ecosystem-based definition. Let us examine a sequence of alternative approaches toward defining and understanding operations and operations management that help understand how commonly adopted perspectives of operations relate to the creation of present sustainability challenges. With the wisdom of Einstein's opening statement of this chapter in mind, the sequence progresses to another level of awareness of operations in ecosystems and intends to enable effective contributions to resolving the sustainability challenges.

1. **The Empirical Organizational perspective:** This perspective is already introduced above and defines operations and operations management in relation to the activities and structures within an organization for which the operations manager holds responsibility.

Let us illustrate this approach by considering operations in two different hospitals. In one hospital, the operations manager is responsible for the operating theaters and all activities taking place in the operating theaters. In the other hospital, the operations manager is responsible for the building, including maintenance and cleaning, the gardens, and the parking facilities. If we seek to develop an understanding of how operations contribute to global sustainability challenges, such context-based differences in the definition of operations are problematic. Should we only consider GHG emissions from the operating theaters for one hospital and all GHG emissions from the building, gardens, and parking lot for the other? This can easily lead to an inconsistent and incomplete measurement of the sustainability of operations and is likely to result in ineffective solutions. For the research purpose at hand, it is necessary to have globally valid definitions for operations and operations management.

2. **The Primary Process perspective:** In this perspective, operations management is the design, control, and improvement of the processes to create the primary products and services of an organization. The corresponding processes, are called primary processes, and operations are subsequently defined as the collection of activities of which the primary processes consist [312].

   The clarity this definition provides in distinguishing operations from other business functions and their corresponding secondary or support processes (such as Human Resources or Finance) has benefits from a management perspective. For the research aims presented above, however, this approach still has disadvantages. Consider, for example, the activities related to the cleaning of the truck owned and used by a manufacturing company to deliver its products to customers. Truck cleaning is not a primary process for such a manufacturing company. However, if the company outsources truck cleaning to a specialized cleaning company whose primary services include truck cleaning, the same set of activities becomes a primary process, and hence truck cleaning now becomes an operation. The same may apply to restaurant services for personnel or to the sterilization of medical equipment in hospitals. Hence, this definition still implies operations to be organization-dependent rather than defined uniformly, consistently, and completely, as is undesirable from a sustainable perspective for the reasons presented above.

3. **The Value Chain perspective:** In part, the shortcomings of the previous primary process-based definition can be resolved by taking a value chain perspective. This perspective considers all activities in an organization as operations, whether part of the primary process or not, and fully recognizes

that many organizations outsource primary and secondary processes to other organizations. Outsourcing happens across national and continental boundaries, and the global value chains in manufacturing and services have become longer while the value added per stage has diminished [20]. Truck cleaning is now considered an operation, whether it is outsourced or not.

New platform business models have even emerged that principally rely on operations provided by other parties. Taking a value chain perspective, the activities of all organizations involved in delivering goods and services to end users are considered operations [225]. Hence, in such a definition, the GHG emissions of taxis contracted by ride-hailing platforms are included even though the drivers are not employees and the taxis are not part of the asset base of the company running the ride-hailing platform. Likewise, pay below the minimum wage, child labor, and other socially non-inclusive practices are included regardless of contract manufacturing practices. This perspective considers all formal value-adding activities contributing to GDP as operations. It includes "the way we work."

Despite its complete inclusion of operations in formal organizations, this perspective may be considered incomplete as it disregards informal operations, as included in "the way we live." If parents bring their children to school by car, instead of outsourcing this activity to a ride-hailing company, this activity still takes place and still impacts sustainability. The same applies when cooking for a hospitalized family member in the absence of (or in preference over) meals provided by the hospital. Moreover, there are many operations at home that impact sustainability, for instance, when using heating, air conditioning, hot water supplies, or when disposing waste. As depicted in Table 1.1, these activities represent 19 percent of GHG emissions and, more importantly, impact other sustainability measures [437]. The operations of life contribute to the much wider informal economy that employs more than half of the global workforce and may represent more than 10 percent of the global GDP [67, 512]. These informal operations are significantly and directly associated with social, economic, and environmental sustainability, and it is thus important to include the management of informal operations as well.

4. **The Human Society perspective:** In pursuit of a definition for sustainable operations management, Kleindorfer et al. [303] required it to encompass "*the set of skills and concepts that allow a company to structure and manage its business processes to obtain competitive returns on its capital assets without sacrificing the legitimate needs of internal and external stakeholders and with due regard for the impact of its operations on people and the environment.*"

This definition includes all business processes and is alignment with the Paris Agreement that it precedes by 10 years [397] as it includes society in the form of external stakeholders. It still limits operations to activities conducted by companies.

If we further broaden this definition to include all operations of work, for instance, the operations of the government and NGOs, and the operations of life, we arrive at a definition that consistently and completely includes all human operations, i.e., all activities conducted and/or controlled by humans. Operations management then correspondingly regards the design, control, and improvement of all human operations. It includes operations that are executed by machines (e.g., computers) or other living species (e.g., horses or bees) under human control.

5. **The Planetary perspective:** The human society perspective still excludes ecosystem services, i.e., the operations conducted by nonhuman organisms beyond the control of humans, such as the carbon captured by the trees of a tropical rain forest. It thus also excludes the operations conducted by machines (such as computers) beyond human control. The latter is likely science fiction at the time of writing, yet a main concern regarding the development of artificial intelligence as part of the 4th industrial revolution. Chapter 11 revisits these developments.

The planetary perspective on operations is the most complete and comprehensive perspective. However, it is problematic from an operations management perspective. If operations management regards the design, control, and improvement of operations, should it regard operations not controlled by humans? Most of these operations appear not to be purposefully designed, controlled, or improved, if at all. The planetary perspective, therefore, appears overly ambitious for the purpose of this research. In pursuit of the research questions on sustainability, we adopt the human society perspective for a definition of operations management and exclude the operations that are not under human control, i.e., the ecosystem services. Building on the previous primary process-based views introduced above, we define [312]:

Definition: Operations management is the design, control, and improvement of human operations.

Restricting the definition to human operations far from excludes considering the influence human operations have on other living species, whether intended, conscious, or unaware. In fact, these interactions, and more specifically their negative impact on nonhuman species and operations, motivate the environmental dimension of sustainability, e.g., in the form of considering planetary boundaries such as (negative impact on) biodiversity [151]. Combining this definition of operations management with the definition of sustainability presented above, we may thus explicitly define:

Definition: Sustainable Operations Management is the design, control, and improvement of human operations to serve the needs of present and future generations of living organisms in an ecosystem, including the economic and social needs of humans living in the ecosystem.

This definition of sustainable operations management explicitly includes the environmental and social inclusion perspectives of operations management, which have received little attention until quite recently [591]. As will become evident in subsequent chapters, the presented broad definition of operations management also responds to the call for "*embracing the multidisciplinary nature of the sustainability challenge*" [163].

The presented definition is not intended to invalidate alternative perspectives on operations management and corresponding definitions. There are many practically relevant operations management questions for which an ecosystem-based perspective may not be most instrumental. However, it may serve to find a different, broad perspective that opens up a broader solution space than offered by commonly adopted approaches to address the sustainability challenges caused by human operations [303]. The severity, urgency, and irreversibility of some of these impacts compel the operations management discipline to adopt new perspectives and "levels of awareness" that best enable them to resolve the sustainability challenges caused by operations.

Each of the subsequent chapters will end with a section devoted to the operations management perspective. Moreover, this perspective is especially relevant in Chapters 10 and 11, which explore the design, control, and improvement of present and future operations to resolve the sustainability challenges.

DOI: 10.1561/9781638282853.ch3

Chapter 3

The Origin of Operations

The honeycomb is made from flowers, and the materials for the wax they gather from the resinous gum of trees, while honey is distilled from dew, and is deposited chiefly at the risings of the constellations or when a rainbow is in the sky.

Aristotle [440]

3.1 In the Beginning

The origin of operations is defined by the emergence of life, of a community of members of living species, in the ecosystem of planet Earth roughly 4 billion years ago (see also Chapter 2) [388, 407].

Much later, 530 million years ago, the supercontinent Gondwana reached its maximum form on planet Earth [378]. It contained the continental mass of all present continents and has greatly influenced the evolution of the planet. Gondwana is depicted in Figure 3.1 [62]. The assembly of Gondwana coincided with the appearance of the first terrestrial plants (such as mosses), and evidence suggests animals living in oceans started visiting land around this time as well. Gondwana provided the environment for seminal terrestrial ecosystems with terrestrial operations.

After its assembly as an initial environment, Gondwana would experience a series of rifting events, breaking the supercontinent apart. The breakup was completed

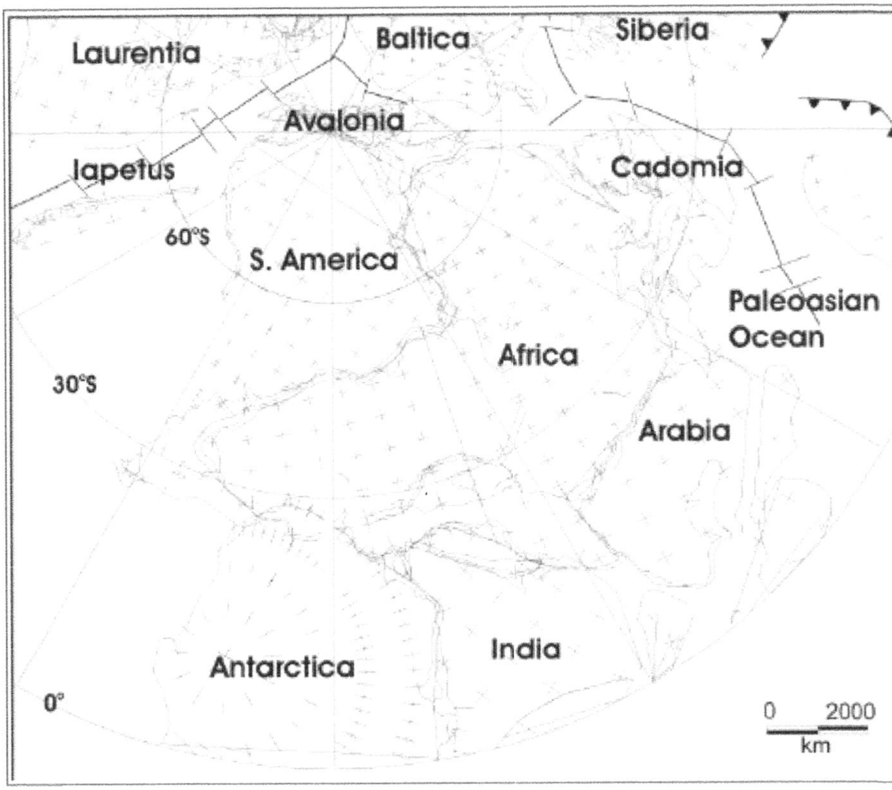

Figure 3.1. Map of Gondwana around 544 million years ago, source [235].

around 100 million years ago. Some animal and plant species from the community living on the supercontinent prior to breaking up thus ended up in multiple newly formed continental ecosystems. This applies, for instance, to some insect species, such as bee species, which appeared 120 million years ago [381]. Bees and the operations of bees will be studied more explicitly below and returned to in various subsequent chapters as we pass through the timeline of operations and its sustainability.

As the parts of Gondwana disconnected over time, land-living species that developed later spread across fewer continents and were therefore absent in ecosystems on other continents. This applied, for instance, to some bee subspecies, such as the honey bee, as covered in more detail below, and to hominin species and their successors such as Homo sapiens, whom we will study from Chapter 4 onward.

While ecosystems thus developed dynamically and new species emerged, other species disappeared because of the physical processes within the ecosystems and external influences. Volcanic activity and meteorite impacts have, for instances, disrupted ecosystems. Such disruptions have resulted in mass extinction events, some of which eradicated up to 50 percent of animal species or 40 percent of plant species living on planet Earth [374]. The majority of species that have lived on planet Earth have become extinct at some point in the past 4 billion years. The current climate change threatens to cause another disruption of ecosystems with such large-scale and irreversible consequences for biodiversity.

The resilience of ecosystems and species living in these ecosystems clearly illustrate Darwin's evolutionary principles of the survival of the fittest [140]. In cases of disruptions, some species managed to responsively cope with the changes in the physical processes and communities of living organisms. On the other hand, species that were unable to generate and raise offspring before passing away under the new circumstances became extinct. In addition to the operations directly related to reproduction, this requires individuals to feed themselves and avoid lethal events before having successfully raised their offspring to become the next reproductive generation.

A Darwinian view of operations thus highlights their primary purposes for the survival of the individual and the species. Ecosystem engineering can then be viewed as the operations of altering the ecosystem to improve the likelihood of survival and successful reproduction. Conversely, the present attention to sustainability is due to the evidence that our current niche construction practices are harmful for the survival of humans and many other living creatures and species. This Darwinian perspective on operations will be evident in the next two sections illustrating the ecosystem engineering operations of bees and beavers and will also be valuable in subsequent chapters addressing humankind and its operations.

3.2 The Industrious Bees

In his foundational biological work on animal life from the 4th century B.C., "*Historia Animalium*," Aristotle considers a large variety of animals and their operations. He pays extensive attention to the "*industrious*" honey bees and describes their operations in detail [440]. The operations of honey bees illustrate the complexity and advanced organizational behavior that nonhuman operations may involve and form a natural anchor for sustainable operations.

Early bees co-evolved with flowering plants for more than 100 million years. The foraging operations of the bees aided the pollination of the plants, and vice versa. All of the early bee species were solitary, as are most current bee species [381, 492]. Solitary female bees live in nests without males, and their offspring leave the nest.

Over many millions of years, some subspecies developed non-solitary ways of living and working. For many of these subspecies, ecosystem dynamics determine whether they operate solitary or in colonies with various forms of social behavior [381, 492].

Some of the subspecies that emerged, among which are the honey bees, always live in colonies. Honey bees (Apis) appeared some 30 million years ago on the connected Eurasian and African continents [519]. The presently common honey bee subspecies, Apis mellifera, might go back a million years [519].

The operations of bees thus have an interesting timeline of their own; they started out to live and work solitarily, and then some subspecies developed toward **eusocial** ways of living and working. Eusociality is defined as [381]:

1. Division of labor (egg layers versus foragers),
2. Cooperation among adult females, and
3. Two generations (mother and daughters) living together.

It is worth noticing that the words "labor" and co-"operation" appear in the definition of eusociality.

Aristotle's qualification of bees as "industrious" particularly refers to the "*workers*," the daughters of the queen, who form the majority of bees in a colony [440]. Their brothers, known as drones, also live in the hive yet are of less interest from an operations perspective. Before turning to the operations of the worker bees, let us, however, briefly address the operations of perhaps the most important inhabitant of a bee hive, the queen.

After the previous queen has left with a swarm of worker bees or has passed away, one or several new virgin queen bees may be born in a colony. The first important operation of any of these virgin queen bees is to kill all other virgin queen bees and become the sole queen of the colony.

For the single candidate queen bee that manages to successfully complete this first operation, the next operation is to fly and mate with multiple drones. This flight will be the only time the queen is fertilized. After fertilization, the queen returns to the colony and will soon be unable to fly. She shifts her focus to the operation of egg-laying.

A queen bee may lay up to 1,500 eggs a day [404]. The queen bee puts each of these eggs in a cell prepared for this purpose by the worker bees. She puts the smaller fertilized eggs that will turn into female worker bees in the smaller cells and the larger unfertilized eggs that will turn into drones in the larger cells. Next to working on her own reproduction, a queen bee "polices" reproductive activity by worker bees through egg eating and aggression [600]. The queen also secretes pheromones to influence (reproductive) behavior of worker bees [612]. While being continuously tended by nursing worker bees, queen bees may be considered to be hard-working themselves.

With the labor of reproduction being primarily allocated to the queen and the drones, the operations of the worker honey bees focus on the provisioning of food and safety. Every worker honey bee turns to corresponding tasks shortly after completing the metamorphosis sequence from egg to larva to pupa to adult bee. During a 2- to 20-week life span, the sets of operations performed by worker honey bees change over time (a phenomenon known as age-related polyethism) [459, 611].

Winston [611] divided worker bee labor into a sequence of sets of operations performed over their lifetimes. The operations of bees gradually advance over the following five stages:

1. Immediately after breaking from the egg, worker bees start cell cleaning and cutting brood.
2. Tending brood and attending to the queen.
3. Comb building, general cleaning, food handling, and other activities inside the hive.
4. Outside tasks such as ventilation, guarding, and orientation flights.
5. Foraging for food, water, and nest construction materials [429].

The transitions between these operational stages trigger physiological adjustments to more effectively execute the next set of operations [611].

While much communication and coordination take place in and around the hive in the first four stages, the communication among foraging bees in the fifth stage is particularly noteworthy. After returning to the hive, foraging bees can communicate the location of rich sources of food, such as pollen of from flowers, by dancing [589]. The round dance is used to communicate about a nearby source of food, while the waggle dance is used for locations further away. The bees communicate both the distance of the food source and the direction relative to the direction of the sun by dancing, as illustrated in Figure 3.2. These dances can be viewed as a form of operations management aimed at controlling and improving the operations.

3.3 The Engineering Beavers

For a second illustration of nonhuman and pre-human operations, let us turn from insects to mammals and, more specifically, to beavers. Beavers have received much scientific attention for being quintessential ecosystem engineers. They have the capability to dramatically alter landscapes and the biodiversity within these landscapes [624].

Through a common ancestor, the extant species of European beaver and North American beaver are both genetically equipped with the operational capability to build dams [325, 499]. Beavers construct primary dams to establish a lodge pond in which they build large lodges that provide safety from terrestrial predators [33]. These primary dams can be maintained over generations and reach considerable

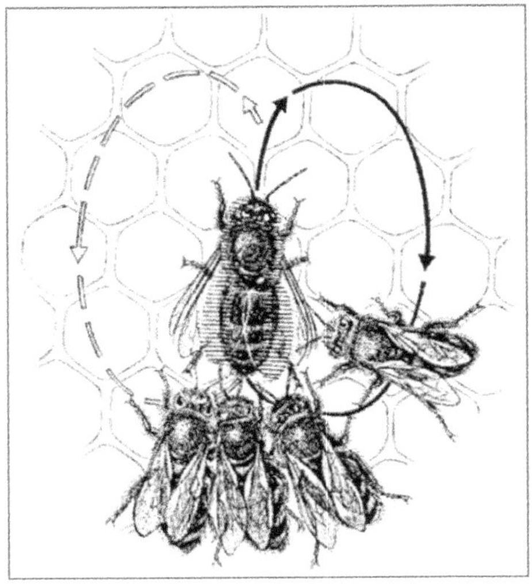

Figure 3.2. Illustration of the waggle dance by which bees communicate the locations of rich food sources, source [589].

Figure 3.3. A beaver lodge in a pond created by a beaver dam, source [395].

lengths. Exceptional cases of dams more than 650 meters long have been reported. Large dams turn land into correspondingly large ponds [497].

The dams are products created by the operations of beavers as a form of nice construction. They establish favorable niches in wetlands, which in turn feed back positively on natural genetic selection for the beavers and other species thriving in the wetlands [326].

Beaver lodges are spacious buildings made of wood and can, in exceptional cases, be several meters high and over 10 meters wide [33]. Access to the lodge is from the bottom and deep enough to remain open throughout a cold winter. This requirement may cause beavers to dig a tunnel if their lodge is on a bank.

Wood cutting (tree harvesting) and swimming are two the main operational capabilities of extant beavers [499]. By building secondary dams and ponds as well as shallow channels extending from these ponds, beavers improve the safety of access to the wood and trees they use as food and building resources [75]. Moreover, the channels improve the efficiency of transportation operations. In view of all these operational capabilities, beavers have long been qualified as engineers and have gained a reputation for being industrious builders [33, 231].

3.4 Sustainability of the Operations of Bees and Beavers

From the next chapter onward, we shall systematically cover all three areas of sustainability: economic development, social inclusion, and environmental protection. In this chapter, the first two areas are not addressed as they refer to human economic activity and social inclusion among humankind. Let us therefore solely consider the operations of bees and beavers from the perspective of environmental protection.

It may appear somewhat odd to examine the environmental protection of these ecosystem services, as they may be argued to form part of the environment that requires protection. However, it will be valuable to view the ecosystem impact of bees and beavers as a reference for later reflections on the ecosystem impact of humans, because the niche construction operations of these species do impact their ecosystems. Moreover, in subsequent chapters, we will explicitly examine the interactions of human operations with the operations of bees and beavers.

As is well known, foraging bees aid in the pollination of plants with flowers rich in pollen. Conversely, these plants enable the foragers to bring in large supplies of food. Indeed, over many million years, bees and flowering plants have co-evalled and promoted each other's sustainability. Through their waggle dances, eusocial bees succeeded in preferentially visiting flowers rich in pollen, thus especially aiding their pollination and giving a relative advantage to corresponding plants over other plants. Conversely, plants with flowers rich in pollen were especially likely to provide bee populations with enough food to accumulate an inventory for seasons in which foraging yields were lower. In many ecosystems, such inventories are needed for a population to survive the winter. Moreover, populations that forage sufficiently can grow large enough to split into two or more populations, thus increasing their long-term survival odds (and illustrating the relevance of the Darwinian perspective on operations).

Obviously, the flourishing of plants with flowers that are preferentially attended by bees has subsequent ecosystem effects. These plants may play a role in the food chains of other species (which is why humans started to keep bees). Moreover, the multiplication and spread of these plants may leave less opportunity for other plant species to survive in the same ecosystems, which in turn may have effects on corresponding food chains.

Negative impacts on the survival of other members of the community of living species in the ecosystem may be marginal for bees but are more extensively researched for beavers. The dams, ponds, and lodges that the engineering beavers create for their own safety and subsistence have considerable impact on the ecosystems in which they live. The impact can be perceived as negative when the beaver ponds lack oxygen, have high carbon concentrations, and thus increase carbon production and emissions. The ponds and wetlands may also provide unfavorable habitats for several pre-existing species [495].

On the positive side, the ecosystem engineering of beavers serves to prevent erosion, store upstream (ground)water, and may improve water quality. These benefits cause the ecosystem to be more favorable for a variety of plants and fish, with increased biodiversity when compared to alternative uses of the same area, such as farming [75].

The above illustrations of the effects of ecosystem engineering show that, regardless of the species practicing ecosystem engineering, ecosystem engineering typically brings a mix of positive and negative effects for present and future generations of living species in the ecosystem. Ecosystem services can create value and destroy value. Subsequent chapters address the effects of the ecosystem engineering operations of humankind.

3.5 An Operations Management Perspective

Regarding the eusocial ways of living and working of the honey bees, we have already noted that the first two of the defining characteristics of eusociality—division of labor and cooperation—clearly refer to operations and operations management. The operations of the eusocial honey bees comprise a complex, coherent, and well-structured set of activities conducted by individual bees as a function of the population. It also involves communication, and learning, and the construction and maintenance of a shared facility to work and live in. It shows how nonhuman species divided labor and cooperated to manage fairly complex operations long before the entrance of humankind on planet Earth.

Honey bees are not unique. There are other eusocial insects, such as bumble bees, some wasp subspecies, and ants. Ants are also well studied and have received considerable attention for their operational effectiveness and their communication through pheromones [284]. In fact, ant colony optimization has become a standard method in the operations research realm that has been applied successfully to solve operations management problems in logistics and manufacturing, which can be modeled as traveling salesman problems [161, 162].

Great apes also practice forms of division of labor and cooperation [372]. Chimpanzees, for instance, have been observed to divide labor among the sexes. Females involved in nurturing offspring are more likely to gather food in quiet, nearby locations, whereas males are more likely to hunt. Cooperation is especially beneficial for chimpanzees when hunting larger animals.

It is also of interest to note that chimpanzees use tools such as wooden sticks for gathering (e.g., for "termite phishing," see Figure 3.4). The operation of these tools requires relatively advanced dexterity involving two hands (or a hand and a foot) [372]. The design, manufacturing, and procurement of such tools is an operation management activity in itself that has no direct reward. Such **indirect reward activities** have been classified as cognitively demanding as they require envisioning an indirect benefit, i.e., a benefit beyond the completion of the current operation.

One might well argue that the construction of beehives, waggle dancing by honey bees, or the construction of dams by beavers are also indirect reward operations and that their rewards are even more cognitively demanding to envision. The capability to conduct these indirect operations, however, is mostly genetically inherited rather than cognitively learned.

Figure 3.4. A chimpanzee phishing for termites with a tool (gathering operations), source [372].

Whether pre-programmed genetically, learned, or obtained through a mix of both, activities such as waggle dancing, manufacturing and use of tools, and collaboration while hunting can be regarded as nonhuman forms of operations management. We note that such nonhuman operations management activities impact sustainability. They impact the prosperity of the population of animals conducting them, the social inclusion within the population, and the ecosystems in which they reside.

While we have defined operations management to relate to the design, control, influencing, and improvement of operations in ecosystems by humans, we may thus recognize that ecosystem services also involve forms of operations management. Operations management behaviors, as, for instance, practiced by greater apes, are akin to hominin forms of operations management on which the capabilities of early humans are founded. Perhaps it is indeed somewhat artificial to distinguish human operations from those conducted and controlled by other species. The remainder, however, mainly focuses on human operations, as these are the operations threatening transgression of the boundaries of a safe operating space. Where of interest, the interrelationships between human and nonhuman operations (management) will be covered as we follow the timeline of operations.

DOI: 10.1561/9781638282853.ch4

Chapter 4

The Operations that Shaped Humankind

The greatest gift that Mother Nature gave to mankind is rapid oxidation, or fire.

Andrew Smith

4.1 Stone Tool Manufacturing and the Operations of Hominins

The African savannas and forests formed a diverse ecosystem when hominin species appeared 5 to 7 million years ago, forming a new branch of the family of the great apes [28, 227]. The earliest hominins fed on a vegetarian diet mostly consisting of raw plants. They lived and operated primarily in trees, which provided safety from terrestrial predators [621].

Different from the other great apes, however, the first hominin species were bipedal; they walked only on their legs when on the ground [452]. This freed up their arms for other operations when on the ground, where they increasingly resided. Corresponding physiological characteristics included relatively long arms and curved fingers in comparison to other great apes, who used them for quadrupedal locomotion on the ground [13]. Adult hominins had a body weight of 30 to 40 kilograms and a brain size of 300 to 350 cc [452].

In comparison to other great apes, the physiological differences in the wrists and fingers made hominins more capable of operating primitive tools such as stones [13]. This operational capability would have a decisive effect on their development.

Chapter 3 already briefly covered the operational ability of (other) great apes to use tools [372]. Present-day chimpanzees are also capable of making more advanced, flexible use of tools for a variety of purposes [249]. However, the arms, wrists, and fingers of chimpanzees have developed in support of tree climbing and quadrupedal motion when on the ground, which is associated with physiological limitations in tool use.

The first use of tools such as stones by hominins may well have been for foraging raw plants, i.e., in gathering operations. Moreover, sharp stone tools have likely started to support the addition of meat to the diet some 2.5 million years ago, when hominins cut carcasses [13]. These scavenging activities can be viewed as a stepping stone toward hunting operations. These physiological and dietary developments co-evolved with a first, unnumbered, industrial revolution: stone tool manufacturing [488].

"To manufacture" means to make by hand, and with their distinct physiology, hominins uniquely developed the operational capability to produce stone tools [380]. The oldest known "industrial complex" for stone tool manufacturing is the 2.5 million-year-old Oldowan industrial complex in the Ethiopian Rift Valley [332]. Figure 4.1 provides an image of an Oldowan-manufactured chopper.

Disregarding developments by other species, such as the transition to eusociality by honey bees, historical evidence thus suggests that a first industrial revolution

Figure 4.1. An Oldowan-manufactured stone chopper (source [54]).

may have taken place 2.5 million years ago in present-day Ethiopia. The differences between the operations in this industrial complex and the operations in industrial complexes of modern times are manifold and perhaps beyond comprehension. Nevertheless, the hominins working in these industrial complexes have been characterized as being highly skilled at consistently producing hand-held stone-flaked tools [13].

The Oldowan tools improved access to energy-rich food sources such as thick-skinned animals, which hominins may have hunted collaboratively. (Collaborative hunting is also practiced by some other great apes [372].) The effectiveness of collaborative hunting for larger animals, thus moving up the food chain hierarchy in the competitive ecosystems of the African savannas, required coordination and favored individuals with better cognitive and physical abilities. Conversely, the higher caloric value of a partially meat-based diet supported larger bodies and brain sizes of up to 800 cc. The physiological evolution of the early hominins was thus closely intertwined with their operations. The bundle of physiological and operational developments taking place at the time of Oldowan manufacturing formed an adaptive threshold in the evolution of humankind [13]. The hands and arms of modern humans resemble those of (some of) the hominin species that adopted Oldowan stone manufacturing.

The genus Homo, which encompasses the species Homo sapiens, humankind, appeared in the African forests and savannas no later than 2.3 million years ago [452]. The early homo species were among the adopters of Oldowan stone tool manufacturing practices.

Early homo species subsequently innovated the manufacturing operations of the stone tool industry around 1.5 million years ago. The workers in these newly arising "Acheulian" industrial complexes were capable of producing hand-held tools of up to 20 centimeters long in pre-designed shapes. The design, manufacturing, and use of these tools required more advanced cognitive abilities, and the brain sizes of the homo species working in Acheulian industrial complexes reached up to 1,200 cc [452]. Acheulian tool manufacturing practices have continued for over a million years, until at least 300,000 years ago [13].

Before advancing, let us briefly reflect on the above from an operations perspective. The operations of many animal species include the use of materials encountered in the environment of the ecosystems in which they live. Like bees and beavers, animals of a wide variety of species forage materials to build nests, hives, homes, dams, et cetera. Moreover, their operations may involve manipulating these materials with their teeth or claws. Such use of materials as a resource or as an input is direct, as the input itself is consumed or utilized in construction. The use of materials as a tool or as a device to assist in the transformation process is indirect and therefore different. As mentioned in the previous chapter, some other great apes practice primitive forms of indirect operations, e.g., when modifying a stick to increase "termite phishing" effectiveness [372].

Nevertheless, the stone tool manufacturing operations of hominins and their successors are a revolutionary, threshold development toward indirectness of operations. Compared to searching and using tools, tool manufacturing is essentially different as it requires the cognitive ability to envision and appreciate that an extensive and time-consuming effort, which is different in nature from hunting or gathering and not directly related to a specific short-term hunting or gathering activity, will pay off because of the prolonged future operation of the tool [249]. The conscious conduct of operations to achieve a desired goal (e.g., obtaining food or shelter) in an indirect manner is cognitively more demanding than direct reward operations. The practice of indirect operations co-evalled with brain size increases.

4.2 Making Fire and the Operations of Early Homo Species

The second main adaptative threshold in the operations of hominins occurring roughly alongside stone tool manufacturing is the use of fire [488]. The history of the use of fire is difficult to reconstruct, as evidence of fire is scarce and difficult to interpret [107, 621].

As was the case with stone tools, fire was not invented by humans. As already mentioned in the opening lines of this chapter, fire, or rapid oxidation, can also be provided as an ecosystem service by Mother Nature. Depending on the season, wildfires occur regularly through natural causes such as lightning strikes in the African savannas and in the African forests, the ecosystems in which hominins have appeared and developed since 5 to 7 million years ago [28]. Early hominins were likely capable of understanding the behavior of fire [28]. The cognitive abilities of early hominins sufficed to interact with fire beyond flight and other fear responses. Over time, this led to the use of fire as a resource in the operations repertoire of various homo species. Let us briefly review the steps taken by homo species to adopt the use of fire in their operations, based on the available evidence, the absence of evidence, and various corresponding scientific interpretations.

Early homo species and their predecessors had developed a familiarity with fire, learned the danger it posed to their safety, and likely witnessed the benefits of fire through their regular exposure to wildfires in the dry season on the African savannas. Fire provides light and warmth at night and keeps predators fearful of fire at a distance [107, 488]. The light and comfort of fire facilitate social gatherings after sunset. It can thus provide extra and joint productive time, for instance, to produce stone tools after utilizing the daylight for foraging. Moreover, plants and animal meat heated by fire provide more easily digestible sources of food with fewer health risks in comparison to non-cooked food [621].

The cooking hypothesis suggests that the consumption of cooked food, which allowed increased calorie intake with fewer challenges, has greatly influenced the development of homo species. It enabled larger body and brain sizes and had major behavioral consequences [621, 622]. Consumption of plants and meat cooked using naturally occurring (uncontrolled) fire and taking advantage of the heat, light, and safety provided by natural fire may well have formed the first steps in the incorporation of fire and rapid oxidation in the operations of early homo species.

It is difficult to establish the earliest use of fire by Homo species beyond taking direct advantage of natural fire provided as an ecosystem service because the earliest of such uses may not have left any evidence and—conversely—encountered evidence of fire may relate to natural fires without use by homo species. The earliest evidence of fires tended by humans may date back to 1.6 million years ago and is associated with early Acheulian activity [28, 488].

From an operations perspective, it is important to appreciate the demands of the transition from enjoying the benefits and effects of naturally occurring wildfires to controlling fire in operations for an indirect reward. The early forms of such functional use must have relied on actively manipulating and maintaining naturally occurring fires, as initiated, for instance, by lightning strikes or spontaneous combustion [488]. African savanna ecosystems with seasons of regular wildfires accommodated the development of such opportunistic, or fortuitous, use [504].

Maintenance and use of fire are advances in operations that require cognitive abilities of anticipatory planning, response inhibition, and future-directed cooperation [565]. The additional efforts to forage feedstock in advance and spend time to keep a fire burning (overnight or while others are hunting) are not worthwhile when unable to envision future use and benefits (in the evening or even the next day). Likewise, for some group members to take on the task of tending the fire instead of going hunting for meat or gathering plants requires to suppress direct reward foraging needs. It requires joint operations management and involves communication, cooperation, and division of work to jointly achieve higher benefits for the individuals involved. It also requires trust and fairness within the group and dealing with theft between groups [622].

As long as the gains made by such an investment of time and effort to learn new abilities and operating modes were uncertain, the required corresponding changes in the ways of working and living were of limited value. Rain or wind could end a fire. Moreover, the inability to transport fire from one hearth to another, as required when a location started to provide insufficient foraging returns, would further limit the value. Hence, hominins faced considerable challenges to the dependable use of fire and its associated benefits. This may explain why such innovations in operations progressed slowly, requiring hundreds of thousands of years while the opportunity continued to present itself seasonally.

The difficulties diminished when acquiring the capability to transport fire. This facilitated bringing fire along from hearth to hearth and the sharing of fire among groups, thus increasing its availability and making it more attractive to rely on fire, even if only seasonally.

A much more decisive innovation in the control of fire was the ability to make fire [504]. The ability to make fire resolved problems faced by homo species in the African savanna in seasons without wildfires or when not in contact with other groups. Moreover, this operational capability is strictly required to enjoy fire in other environments that lack regular occurrences of wildfires for fortuitous use. Hence, the habitual use of fire by homo species 800,000 years ago in the Jordan Rift Valley implies they were able to make fire [11].

The cognitive and physiological demands of creating fire go beyond the demands for stone tool manufacturing. Some present-day hunter-gatherers manufacture fire by rotating a hard wood drill into a soft wood base, as depicted in Figure 4.2 [107].

Figure 4.2. Present-day hunter-gatherers of the Hadza tribe making fire (source Science Photo).

This activity requires considerable skill and effort and is essentially different from the processes by which naturally occurring fires are controlled [334].

The significant cognitive and physiological barriers to making fire might explain why it took hominin species millions of years to envision and develop this operational capability. Moreover, it may form an explanation for the slow spread of this valuable innovative operational practice. Indeed, other early evidence of human-made fire is scarce, from far away, and mostly from much later. There is evidence of habitual use in China from around 700,000 years ago, a second source from the Eastern Mediterranean region from 400,000 years ago, and the first evidence from Europe from 370,000 years ago [488, 565]. As older existing sites evidencing Acheulian activity in Europe and Asia lack evidence of the use of fire, the spread of Homo erectus to these colder continents, which dates back to over a million years, likely happened without access to fire.

With and without fire, the operations and physiology of homo species advanced, and brain size increased from 500 to 1,250 cc for some subspecies over time [452]. Homo sapiens, humankind, with an adult brain size of roughly 1,300 to 1,400 cc, appeared in Africa around 300,000 years ago [391, 439]. Its relatively large brain requires more energy and, therefore, an increased calorie intake. The large brain size is also associated with increased cognitive abilities that facilitate communication and coordination and, thus, more complex operations to obtain and prepare food and achieve other goals.

4.3 The Operations of First Humans: Hunting and Gathering

The late Acheulian Homo sapiens populations living around 300,000 years ago lived without homes or clothes and likely had only relatively recently acquired skills to manufacture fire. The same is true for Neanderthalers and other homo species living at the time.

Fire was not only used to heat food but more generally supported hunting and gathering operations. It served to improve tools beyond the relative simplicity of hand-held stone tools. Fire was used, for instance, to harden the points of wooden spears. This innovative manufacturing process was complemented by the use of stone tools to produce spears [349, 409]. The archeological evidence available from this period of time importantly consists of tools supporting the operations after which these prehistoric populations are named: hunting and gathering.

Early use of spears may date back to Africa no later than 500,000 years ago and, for instance, have spread to Europe 300,000 years ago. The use of spears, whether thrust or thrown, and later as arrows or other projectile tools, broadened the subsistence opportunities and improved the safety of the hunters [409]. Moreover, spears and similar tools may have served to provide protection and be used for other purposes, including warfare [349].

Over time, and depending on context, the hunter-gatherers made a variety of further innovations in the design of spears and the spear production process. These innovations included the assembly of spears from different components through a sequence of production steps [349]. Some of such composite spears had a stone tip on a wooden shaft, sometimes with the application of adhesives to improve robustness.

Similar composite designs and assembly techniques were developed for other tools, such as axes and hammers. These more complex additive manufacturing processes formed a substantial technological advancement over the subtractive manufacturing operations practiced in Aecheulian industrial complexes. The new assembly operations conceptually and practically extended the distance formed between the activities in the process of manufacturing operations (e.g., procure materials, manufacture components, assemble, hunt, butcher, make fire, cook, et cetera) and the reward of the operation, i.e., nutrition. It thus co-evolved with further cognitive advancement [249].

It is rather unclear when these assembly innovations developed, and it is likely that there was considerable regional variation in these developments [349]. Stone tips found in Africa date back to around 100,000 years ago [409]. The earliest indirect evidence of more advanced equipment, such as a spear thrower or arrow and bow, is from Africa and dates from 64,000 years ago, while the earliest direct evidence is from Europe and much more recent, from 17 to 18 thousand years ago [409]. There is consensus that the hunter-gatherers of the species Homo sapiens, humans, uniquely operated arrows and bows (see Figure 4.3 [349]).

After earlier "Out of Africa" migrations by various hominin and homo species, Homo sapiens' first "permanent" migrations to other continents started around 100,000 years ago. Through hunting and gathering, populations of Homo Sapiens entered new ecosystems in the Levant, the Arabian peninsula, and the southern Asian coasts. This migration reached Australia around 50,000 years ago when sea levels were much lower than present-day sea levels [244]. The migration of Homo sapiens to Europe dates back to 48,000 years ago [268].

Humankind likely entered the Americas in two waves. A first wave occurred around 30,000 years ago, and then again around 14,000 years ago [50]. Humans entered and settled in the most southern part of America 11,000 years ago [465]. In 1520, Magelhaes would give the area the name that is most commonly used at present, Tierra del Fuego, Land of Fire, after observing the fires the resident hunters and gatherers kept lit at night.

Figure 4.3. Early human hunter with arrow and bow—Cederberg Mountains, South Africa (source Science Photos).

4.4 Sustainability of Hunting and Gathering Operations

4.4.1 Economic Development

Through the formal and monetary lenses of GDP and per capita GDP, the economic activity of hunter-gatherers is hard to observe or measure. More so, as these nomads had no homes and no countries to allocate their "domestic" production to. Nevertheless, the more than 5 million years of development in hunting and gathering operations on planet Earth covered above show that gross production per capita increased considerably. Homo sapiens, with its adult body weight exceeding 60 kilograms and brain size of 1,300 to 1,400 cc, consumed and therefore necessarily produced much more energy (per capita) than the initial hominins with body weights below 40 kilograms and brain sizes of 300 to 350 cc.

Additionally, the value of global gross production must have increased considerably since the first hominins and since the first humans because of population growth. Early humans most likely lived in small, rather isolated, and separately evolving populations for more than 100,000 years before dispersing out of Africa. By the time of the agricultural revolution (covered in the next chapter), the global human population had likely increased to one or several millions [52, 302]. The advances in operations such as stone tool manufacturing and the use of fire have promoted this dispersal and population growth and have thus been the main drivers of the corresponding growth in global GDP and in global per capita GDP [52].

4.4.2 Social Inclusion

Archeological records of the mobile hunter-gatherer populations obviously provide limited evidence about their social inclusiveness. It has been observed, however, that the possibilities for personal accumulation of property were limited in the absence of a sedentary environment to buffer materials, tools, or food. Sharing among group members (and even among groups) serves as a mechanism to protect individuals and families against foraging shortages. Such sharing is common among extant hunter-gatherer populations and is considered to be more effective than sedentism or non-sharing in the unpredictable ecosystems in which hominins and early humans operated [95, 246, 342]. Moreover, extant non-sedentary hunter-gatherer populations still exhibit continual socialization against economic, social, and political inequality [95]. Hence, original hunter-gatherer populations likely operated cooperatively and were likely relatively socially inclusive with respect to their own population members.

4.4.3 Environmental Protection

The environmental impact of the advancements of human operations was relatively modest in Africa, where various hominin species and—in a later stage—homo species gradually co-evolved with other species in the local ecosystems over millions of years.

For instance, as homo species developed tools and fire, they became more effective at foraging honey. They used sticks as tools to approach honeycombs in beehives and collect honey, which has the highest energy density of all natural foods [363]. These methods were and are also practiced by other great apes. Later, more advanced honey-collecting operations involved axes to climb trees and break the beehives. As beehives are often fiercely protected by the stinging bees (see Chapter 3), honey-hunting operations started to include the use of fire and smoke, which reduced the defensive activity of the honey bees [363]. The same operations are still practiced by contemporary hunter-gatherers such as the Hadza, for which honey forms an important diet constituent, especially in seasons in which animal hunting is less effective [363].

The effectiveness of honey collection by the Hadza and other contemporary hunter-gatherers is aided by the honeyguide bird, which guides them toward beehives that are relatively rich in honey [536, 615]. These collaborative operations of the honeyguide bird and hunter-gatherers are mutually beneficial, as the honeyguide bird subsequently feeds on the leftovers of the honeycombs made accessible by humans.

Humankind and honeybirds are just two of many more species foraging honey in the African ecosystems in which these species and the honey bees operate. As a species, honey bees have not been threatened by advances in human hunting and gathering operations. Even after the introduction of new operational capabilities such as stone tool manufacturing and making fire, the honey bee species continued to operate and provide ecosystem services in the ecosystems in which bees and humans habituated jointly. This also appears to have been the case for most other animals in Africa affected by the advancement in operational effectiveness of homo species [288]. In Africa, the gradual advancements of (pre-)human operations over millions of years provided most other members of the community of living species in the ecosystem with enough time to adapt and co-evolve.

The migration of humans to other continents had a much more disruptive effect on the ecosystems they entered and has commonly been associated with the extinctions of many species and subspecies, especially of larger mammals that were particularly attractive to hunt. Outside of Africa, all mammal species with adult body weights over 1,000 kilograms became extinct after the entrance of Homo sapiens (see Figure 4.4, [306]). These and other extinctions of large and slow-breeding animals were more profound on continents further away from Africa. In Australia, none of the species with an adult body weight of 100 kilograms or more survived, and all these extinctions are attributed to human operations [306]. The dispersion of humans and the increasing effectiveness of human hunting operations have translated into significant biodiversity loss.

The extinctions of large mammal species in turn had considerable and sometimes devastating consequences for other living organisms in the corresponding ecosystems, such as plants that co-evolved with certain animal species [288]. As a result of large mammal extinctions, forest composition changed, and the extinctions may have contributed to forests becoming denser [224]. Denser forests provided more fuel for natural wildfires and likely contributed to increased fire frequencies and facilitated the spread of wildfires [224]. While other hunting and gathering homo species, such as Neanderthalers, may have contributed to these developments, all these other homo species have gone extinct in the process as well. The dispersal of humankind across planet Earth, with its advancements in operations, has resulted in irreversible changes to the ecosystems it entered and meant the end of the planet as a safe operating space for many species, among them all other homo species.

In the process, humans more frequently entered the ecosystems engineered by beavers. The hunting practices of the human population living in North America after the initial migrations into the continent more than 10,000 years ago had

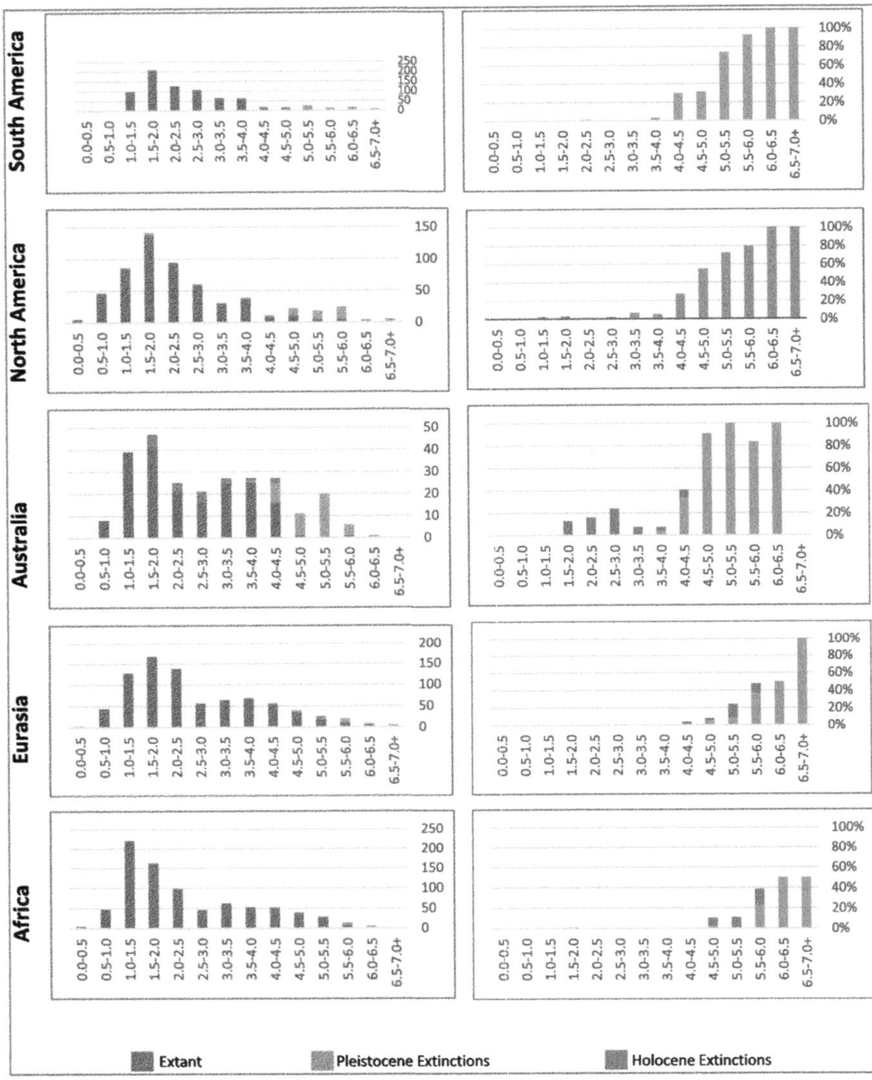

Figure 4.4. Mammal species extinctions per body weight and per continent in absolute numbers (left) and in relative numbers (right). Body weights (horizontal axis) are in ^{10}log, e.g., the category 1.0–1.5 includes mammals with body weights from 10 to 32 grams. The data are from Table S6 in [527] and updated Figure 1 in [306]. Some of the extinctions in red (Holocene) may have occurred after (partially) transitioning from hunting and gathering to agriculture.

little impact on the beaver-engineered ecosystems and its beaver population (Castor Canadensis). When Europeans arrived in North America around 500 years before the present, the North American beaver was highly prevalent throughout large parts of North America, and their ecosystem services and engineering practices importantly influenced a large part of the continent.

The European immigrants actively procured beaver furs through trade with the indigenous population. It would not be long, however, before they procured or conquered land and turned to large-scale beaver hunting [579]. Beaver hunting operations were particularly intense during an era known as "the little ice age" in the early 17th century, which generated huge beaver fur demand from Europe. Dutch immigrants exported 80,000 beaver furs from their modestly sized North American province in the year 1671 alone [497]. Altogether, the beaver hunting operations of the British, French, Dutch, and others may have killed as many as 50 million North American beavers over the next centuries, possibly affecting the global climate because of the reduced carbon emissions from beaver-maintained wetlands [579]. Moreover, beaver trade by various European countries catalyzed social conflict among indigenous hunting populations, resulting in hostilities known as the beaver wars [538].

The North American beaver populations had been marginalized by the turn of the 20th century. This near-extinction of beavers ended the beaver hunting and fur trade operations [231]. By then, the virtual elimination of the North American beavers had significantly altered watersheds, landscapes, and the occurrence of a variety of other plant and animal species in the ecosystems previously engineered by beavers.

The Eurasian beaver (Castor fiber), once widespread on the continent from East Siberia to Portugal, was similarly overhunted by humans and was close to extinction as well by the turn of the 20th century. At the time, the European beaver population had been reduced to some 1,200 beavers, living in eight small and isolated populations [405]. Beavers had, for instance, been hunted to extinction in Great Britain.

By contrast, humans introduced 25 pairs of North American beavers to the aforementioned region of Tierra del Fuego in 1946 with the purpose of commencing fur production and trade operations [19]. Without (other) natural enemies, this beaver population grew rapidly, and their ecosystem engineering operations severely impacted the ecosystems of Tierra del Fuego. The beaver "*invasion*" thus brought "*the greatest modifications to this landscape*" since humankind entered these ecosystems [19, 261]. The areas entered and subsequently abandoned by beavers (after the trees were harvested by the beavers and the ecosystem no longer provided sufficient subsistence) have experienced difficulties recovering even after several decades [435].

Altogether, we may conclude that even in comparison to the effective ecosystem engineering beavers, hunting and gathering humans stood out for their impact on ecosystems while creating their own niches to live and work in [550]. It appears, however, that the resulting (near) extinctions of other species and subsequent ecosystem impacts have often been unintentional consequences of human operations rather than consciously planned ecosystem engineering operations.

Along the way, hunter-gatherers also developed intentional ecosystem management operations, particularly through the use of fire [224]. Early African evidence regarding the use of fire to manage land dates back as much as 85,000 years [556]. These fires were lit and managed to intentionally impact the fertility of the land and the composition of plant and animal species.

4.5 Operations Management Perspectives

Let us recall that we have defined Operations Management as the design, control, influencing, and improvement of human operations and ecosystems. The words design, control, influencing, and improvement refer to conscious efforts to affect operations. As we have observed in Chaper 3, human operations management practices have evolved from pre-human practices, as, for instance, practiced in collaborative hunting by large apes. The Oldowan and Aecheulian stone manufacturing operations by hominins, who predated Homo sapiens by millions of years, required the design and control of processes that consistently produced hand-held stone flaked tools [13]. The subsequent transition from Oldowan to Aecheulian manufacturing techniques demonstrates a conscious, managed effort to improve processes.

The design of a process to make fire is a particularly noteworthy accomplishment. The process is completely different from processes to control and maintain naturally occurring fires and requires considerable skill. It is very imaginative and cognitively demanding. The design and control of the operation to make fire had been accomplished and spread across various continents long before the arrival of Homo sapiens. While there is no direct tangible evidence of the trial and error iterations to design operations for making fire, there is evidence of man-made fire that demonstrates that operations management by Homo sapiens, by humans, builds on many millions of years of preceding advancements.

4.5.1 Operations Management for More Effective Ecosystem Engineering

Many of the operations management efforts to improve operations came in the form of new, more complex, designs of sequences of activities that were more difficult to control and therefore to learn. They developed gradually and incrementally over hundreds of thousands of years, as in the cases of stone tool manufacturing and making fire. These evolutionary advancements in operations have co-evalled with increases in brain size and cognitive abilities. Over time, such gradual advancements sometimes had threshold effects on the development of humankind. On hindsight, the evolution toward a threshold may seem like a revolution in operations once the threshold had been reached.

The increasing complexity of operations increased the distance between the activities conducted and the desired goal (or reward). The complexity caused operations to become less direct. The use of fire to manage land is perhaps the most illustrative example of such indirect reward operating models. The reward of this operation is hoped to materialize one or more seasons later. It forms an important development in operations management and very explicitly relates to human ecosystem engineering. In this case, ecosystem engineering takes the form of a delayed return operating model in which parts of the ecosystem are burned down. This may reduce short-term foraging effectiveness in the area burned down, yet it has the desired goal of obtaining better long-term hunting and gathering yields in comparison to direct hand-to-mouth hunting and gathering.

From a sustainability perspective, some of the innovative human operations management practices yielded economic growth for the early humans while negatively impacting the livelihoods of other species with whom they shared their ecosystem. Homo sapiens entered the Eurasian ecosystem of the Neanderthalers less than 50,000 years ago and may well have been an important factor in the extinction of the Neanderthalers less than 20,000 years later.

Operations management activities that deploy the use of fire to burn land and control foraging returns have been practiced on various continents, including Africa, Europe, and Australia [194]. This ancient operating model has continued to be practiced over the timeline of human operations until today. As we shall examine more closely in Chapters 10 and 11, resulting deforestation causes reduced effectiveness of a variety of ecosystem services, including carbon capture and sequestering, whereas fires emit GHGs. One of the nine boundaries of a safe operating space for planet Earth explicitly regards forestation.

The use of fire by hunter-gatherers to manage land promotes the yields of natural crops and their suitability for herding. It thus helps to gain more efficient and safer access to subsistence. These innovative operations may well have been precursors to the agricultural revolution covered in Chapter 5.

4.5.2 Specialization and the Division of Labor

An important area of advancement in operations management made by hominins and early humans is in the area of specialization and division of labor. Division of labor requires coordination and collaboration. For instance, we have already seen that to keep a fire burning, one person may stay at the camp to tend the fire while others go hunting and gathering. Moreover, the hunting operations may be allocated to other individuals than the gathering operations. As was the case for chimpanzees, differences between the sexes can play a role in these divisions of labor. The evolution favors pregnant females and females with juveniles who practice safe gathering operations over those exposed to the dangerous dynamics of hunting large animals with hand-held tools [246].

However, the somewhat stereotypical division of labor in which men hunt and women gather is not generally valid for present-day hunter-gatherers and less so for early hominins [318, 431]. There is, for instance, little evidence to suggest that Neanderthalers divided labor as pronouncedly as many present-day hunter-gatherers. Rather, it seemed that the scope of operations of the Neanderthalers was rather narrow and that all group members primarily participated in the operations of hunting large game [318]. The varying foraging returns and the risks to personal health were the main disadvantages of this narrowly scoped large game hunting operating model of the Neanderthalers. This likely caused Neanderthaler populations to be vulnerable, remain small, and altogether have low population densities (per square kilometer) [318].

Starting from 100,000 years ago, some of the African Homo sapiens populations developed an operating model based on a much broader set of activities. In addition to hunting large game, they more regularly hunted smaller game. Moreover, the gathering of vegetables, fruits, nuts, et cetera provided for a large part of their calorie intake. The new and more broadly scoped model, in which operations were managed based on division of labor, yielded more reliable foraging returns at lower health risks. These operations management methods provided an evolutionary advantage.

Over time, the Homo sapiens populations that had adopted this model successfully dispersed from Africa across the globe and were able to maintain higher population densities while (re)designing their operations to fit newly encountered ecosystems (e.g., at higher latitudes) [318, 431]. As these populations spread and reached other continents, the set of activities broadened further and started to include the manufacturing and use of tools for non-foraging activities such as the assembly of clothes [318]. This wider set of activities, for instance, facilitated children and pregnant women to conduct operations that were different from the operations of large game hunting, which were mostly conducted by men in extant hunter-gatherer communities [246]. The diversity of tasks and prolonged effort needed to acquire the corresponding specialized skills were essential elements of the division of labor practiced in the operating model of Homo sapiens [246, 318].

DOI: 10.1561/9781638282853.ch5

Chapter 5

Food Production in Sedentary Niches

53. If any one be too lazy to keep his dam in proper condition, and does not so keep it; if then the dam breaks and all the fields be flooded, then shall he in whose dam the break occurred be sold for money and the money shall replace the corn which he has caused to be ruined.

54. If he be not able to replace the corn, then he and his possessions shall be divided among the farmers whose corn he has flooded.

Law Code of Hammurabi [298]

5.1 First Farmers

As climates changed and various glacial periods occurred over the period from 100,000 to 10,000 years ago, humankind spread across the continents of planet Earth [143]. The latest glacial maximum was reached around 25,000 years ago, after which climates continued to be variable. Temperatures tended to increase until a warmer and more stable climate was reached in important human habitats around 11,500 years ago [53].

During this climate transition, humans started to advance their operations beyond hunting and gathering other species with which they shared their ecosystem and, for instance, experimented with the domestication of plants since around

20,000 years ago [53, 634]. These experiments included the propagation and protection of plants and early attempts at cultivation [53]. Across the Near East, this cultivation involved the planting, growing, and harvesting of Einkorn, Emmer, and pulses.

11,500 years ago, humans had succeeded in domesticating crops, thus marking the operational transition from hunting and gathering toward agriculture [634]. Similar, apparently independent agricultural revolutions have occurred in other locations—some in later time periods—such as in the Indus Valley, in China, in New Guinea, and in the central Andes [53].

The adoption of agriculture is a transition to ways of working and living that are essentially different from non-sedentary, nomadic hunting and gathering operations. Especially in colder and less reliable climates, hunters and gatherers typically moved from camp to camp in large and dynamic foraging areas in groups of several families, perhaps a few dozen individuals. The population density may have reached one or two persons per square kilometer [53, 494]. Displacements within large and dynamic areas make it difficult to grow and protect crops. Conversely, crops had to be reliable and sufficient for settlement to become an attractive alternative. Moreover, hunting would have to be sufficiently effective when settled or be (partially) substituted, for instance, by produce from domesticated animals.

In the Near East, the domestication of plants co-evalled with a development in which hunting operations increasingly happened within a smaller area around the location of crop cultivation. As several species of large animals became extinct and others less prevalent, especially in human habitats, hunters necessarily shifted away from dependence on large animals or adjusted their hunting operations otherwise [635]. This could, for instance, take the form of hunting male animals and sparing reproductive females. Over time, such operational innovations developed into managing the movement, feeding, and reproduction of populations of animals, i.e., into herding [635].

The transition toward domestication of animals, including pastoral practices, has not accompanied the domestication of plants by humankind on all continents. For instance, early North American farmers continued hunting instead of switching to the domestication of animals [53]. Some extant hunter-gatherers practice hybrid operating modes as well. Fishing operations can replace or complement the hunting and farming of animals in many ecosystems as well.

The domestication of animals such as sheep, goats, cattle, and pigs developed together with the domestication of crops in the Near East [634], establishing a transition toward fully fledged farming 11,500 years before present. Instead of foraging from the ecosystems in which they moved around, these first farmers engineered the ecosystems in which they operated and adopted a domestic way of living. After almost ten thousand years of experimentation, they managed to change their operating model from a predominantly direct return model, which largely views the ecosystem as exogenous, to a **delayed return model** that enabled them to live and work in a human-created niche. The transitions toward small villages of farmers that had gradually yet fully adopted the agricultural operating model were seldom reversed [53, 586, 634].

The dissemination of agriculture to other locations was neither immediate nor full. In fact, hunting and gathering persists as the main mode of operation for relatively small human populations on various continents until today. Rapid changes in climate have not only enabled agricultural revolutions but also served as barriers to their sustainability. In Mesopotamia, for instance, a partial return to colder and more arid climatic conditions that occurred around 8200 before present resulted in the abandonment of villages and appears to have caused part of the population of first farmers to move westward to Eastern Europe, where first farming practices date back to this period [367, 599]. Rapid climate changes are also viewed as a major cause of the collapse of agricultural civilizations in Mesopotamia and elsewhere at later times, such as the collapse of the classic Maya civilization [148].

As time passed, each of the various agricultural operating models that had emerged at various locations across the globe would spread to amenable terrestrial areas of planet Earth. Less than a thousand years after the initial development of agriculture in the Near East, domesticated crops were introduced to the island of Cyprus, 70 kilometers off the coast [586]. This implies humans have also developed the operational skills to construct boats and navigate long sea distances.

Over the next 5,000 years, transportation by boat across seas and following rivers upstream would also form a main modality to facilitate the spread of Near East agricultural practices across Europe [494]. Likewise, the farming practices originally developed in locations in present-day China would spread across Asia, e.g., to Korea and Japan [53].

The sedentary operations of farmers commonly resulted in increased fertility rates. While mortality rates also increased, for instance, as a result of the higher prevalence of infectious diseases associated with sedentary ways of life, the net effect resulted in population growth [64]. The agricultural operations enabled to sustain a higher population density.

The population growth also drove migration flows into lands previously not inhabited by humankind and into habitats of hunter-gatherers [64]. For many existing hunter-gatherer populations in areas conducive to agriculture, this meant exposure to growing numbers of agricultural migrants. Genetic evidence suggests that the population practicing hunting and gathering operations almost entirely disappeared from Southern Europe as a result of these migration flows yet has continued to exist in Northern Europe, where indigenous hunter-gatherers adopted the new agricultural ways of working and living [449, 494].

5.2 Advances in Operations—The Invention of the Wheel

The development and spread of agricultural operations took many thousands of years, perhaps suggesting that the initial practices provided limited progress, even after a more favorable climate stabilized. Limited initial effectiveness may have implied that agriculture was only considered at times when hunting and gathering yielded poor returns [58]. A number of innovations stimulated the adoption as they improved the relative advantage agricultural operations provided over hunting and gathering. These innovations included the selection and manipulation of plant species and the use of tillage tools, such as ploughs [58, 634].

Making pottery, a set of operations initially invented by hunter-gatherers, was valuable for early farmers as pottery facilitated the storage of seasonal food produced [290]. The support pottery provided for the delayed return agricultural operating model was further promoted by the construction of houses and other community buildings in agricultural settlements to keep inventory [58, 586]. The use of pottery thus spread with the migration of farmers. The invention of the pottery wheel around 6,000 years ago—which indeed marks the invention of the wheel—greatly benefited pottery manufacturing operations. Figure 5.1 illustrates an ancient pottery wheel [493].

Near East inventions closely related to the invention of the pottery wheel are the brick and the use of two wheels and an axle more than 5,000 years ago [84]. The wheels and axes were used with ploughs and for carts to transport agricultural produce. The domestication of horses and donkeys and their deployment for transportation also date back to 6,000 years ago [635]. This marks the first use of animal operations under the control of humans beyond the direct production of meat or milk. Over time, humans living in the niches created by the agricultural revolution invented many subsequent devices and advanced operations in ways that have remained in practice to date.

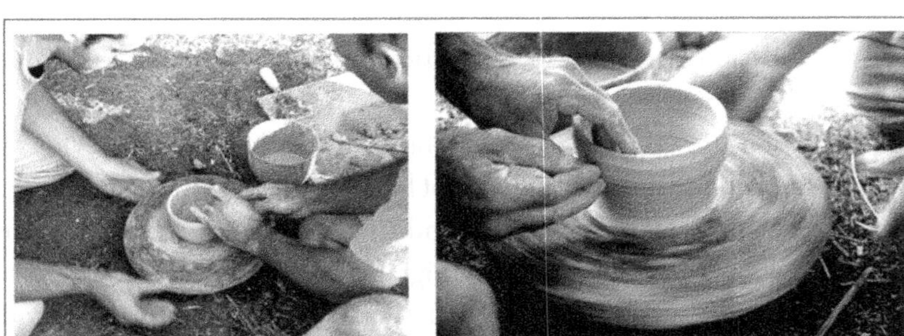

Figure 5.1. Experiment showing the use of a pottery wheel from around 4500 before present, excavated in the Levant. It consists of a small bottom disk (not visible) and a larger upper disk, which is operated by an assistant as the potter crafts (source [493]).

5.3 Metals, Mining, and Melting

The use of metals by homo species dates back at least 40,000 years and thus predates the agricultural revolution. The earliest use of metals by humankind for pigments appeared to relate to spiritual and religious purposes rather than to hunting and gathering operations [241, 483]. The first uses of metal objects likely included adornment and were made from meteoric and "native" metal encountered on the surface of planet Earth [131]. These metals were not the result of mining operations.

The earliest evidence of (open pit) mining of metal ores and subsequent roasting and reduction to smelt the metal regards copper and is from at least 8500 years ago in present-day Anatolia (Turkey) and northern Iraq [131]. The production of metal from mined ore was first practiced by agricultural populations. The innovation of copper smelting in crucibles likely spread from these Near East locations across all of the Eurasian continent and into Africa, in particular Egypt. Mining and copper smelting also developed independently in the Americas, albeit much more recently [181]. A variety of local mining, smelting, and smithing practices arose as communities acquired the metallurgical capabilities to manufacture tools and weapons such as axes, knives, and daggers, as well as objects for adornment and ritual use from metals such as copper and gold [482, 483].

As illustrated in Figure 5.2, the earliest mines were still operated with stone-age resources such as tools made of stone and with the use of fire. The mining processes and metallurgical manufacturing processes would soon advance considerably. Early

Figure 5.2. Ancient copper mine in present-day Iran, with two overviews (photos a and b), two-one man digging spots (c and d), photo c shows copper veins, and a set of stone mining tools (e) (source [402]).

craftsmen developed crucibles and were able to control the temperature and oxygen of a carbon-fueled fire at various temperatures as needed, including temperatures above 1,000 degree celsius to smelt copper ore [483]. These advancements were necessary as the temperatures created by open, wood-fueled fire (bonfire) are insufficient for copper smelting and melting [482]. Further innovations in metal manufacturing operations included the design and construction of furnaces to better control the processing and separation of the resulting metals [483].

Important subsequent process innovations were the processing of ores that contained multiple metals and the deliberate addition of other metals when smelting copper [229]. These innovations facilitated the ease of the smelting process and the casting of objects, as well as product quality. A variety of metals were accidentally or deliberately alloyed with copper, among which nickel and tin naturally co-occur with copper [181]. The alloying of copper and tin yielded bronze, an innovation in manufacturing operations that ushered in the Bronze Age [229, 483]. Hence, we may view that another, more "industrial," revolution occurred soon after the agricultural revolution.

Bronze was strong enough to form the base material for a variety of operating tools. Bronze with a high tin content was, for instance, used to cast swords and other weapons [229]. Later, forging and stamping of bronze developed to produce armor protecting the body and head. Bronze tools such as lancets were produced for use by surgeons, for instance, to operate eyes, as mentioned in the code of Hammurabi around 4,000 years before present [1]. Around the same time, the first bronze musical instruments were produced in China [229]. There is evidence that bronze was used soon after for household operations, e.g., in the form of bowls and other household utensils [229].

Copper ores often also contain iron, and just like copper smelting, iron processing and smelting likely first developed in Anatolia from 4,000 years before present onward [181]. Initial iron objects, likely produced from meteorite iron, appear to have been highly valued. Nevertheless, it lasted around a thousand years before iron mining and manufacturing reached larger scales [181]. As iron is more common than copper and the quality of well-manufactured iron tools became equal to or superior to that of bronze, tools, iron increasingly replaced bronze, and the Iron Age succeeded the Bronze Age.

5.4 Urbanization and Civilization

Advancements in agricultural operations accumulated over many centuries and enabled higher and more reliable yields from farming operations. As a result, fewer and fewer of the inhabitants of the farmer villages were required for the production of food. Agricultural yields nevertheless continued to increase and could exceed the subsistence needs of the villages. These advancements in operations facilitated agricultural settlements to grow their number of inhabitants. Some villages became cities as urbanization intensified 6,000 years before present in Mesopotamia. Among the first cities were Ur, Uruk, Tell Brak, Jericho, and, somewhat later, Babylon. The inhabitants of these cities developed and adopted a variety of operations other than agriculture and mining. The professions of smith and surgeon have already been mentioned. The further division of labor in early cities additionally saw professions such as bakers, leather workers, pottery makers, et cetera. This specialization facilitated economies of scale and promoted further economic growth.

Division of labor and specialization distanced operations further from the communal practices of hunter-gatherers and early agricultural communities toward private practices. Private practice was accompanied by private property and the possibility to accumulate capital in the form of land, seeds, animals, devices, buildings, inventory, objects for adornment, et cetera. Indeed, the products and services created by specialized operations were traded and resulted in the accumulation of capital goods for some population members and families, also in the form of precious metals and—at a later stage—coins.

5.5 Transport, Trade, and Supply Chains

The agricultural communities and civilizations able to produce more food than needed for their own subsistence or to produce artisan products such as pottery beyond their own needs developed trading practices. They sold their production surpluses to other communities in exchange for other goods. Moreover, there is substantial evidence of trade with neighboring hunter-gatherers from ancient times until to day [258]. These trading practices developed further with advancements in transportation and gave rise to the first supply chains.

Mesopotamian cities operated trade networks with Anatolia to procure copper and bronze 1,000 kilometers and more upstream the Euphrates and Tigris since at least 5,000 years before present [414]. In turn, the metal trade networks of Anatolia extended to the Mediterranean region from the Levant to Greece in the next millennium [502]. As there are no tin mines in the Near East or Anatolia, the tin supply chains of the Bronze Age reached eastward to Afghanistan and beyond [620]. The network of land routes would come to connect China and the Near East during the Bronze Age, and sea routes to India were established concurrently [434]. The westward tin supply chains of the Bronze Age reached Britain, and the amber supply chains even reached the Baltic states [620].

Trade with remote locations would initially happen indirectly. Instead of traveling the full distance from the Near East to the Baltic states or China in person, a network of trading hubs connected the full length of these supply chains in both directions [434, 620]. Early Mesopotamian city states, however, already operated networks of trading posts reaching from Anatolia to Bahrain since 4,000 years before present [390]. Such practices were further expanded by Phoenician traders, who gained control over Mediterranean trade networks—with close naval support—and land trading routes to Asia [390]. These trading operations extended to production operations when the Phoenicians set up mining activities, especially for silver, through settlements on the southern European coast in Spain, Sicily, and Sardinia [390]. The continuous supply of tin through these global supply chains may explain the slow uptake of iron as an alternative to bronze (made of copper and tin) in the Near East, despite the availability and accessibility of iron ore from nearby Anatolia. Supply chain management may already have had a strategic impact on business 4,000 years before present.

5.6 Sustainability of the Agricultural Revolution

5.6.1 Economic Development

The transition in operations from hunting and gathering to agriculture has driven tremendous economic development. Indeed, it marks the beginning of the era in which the word domestic applies. Together with the development of coins and trade, it is from this era that GDP can be financially measured. First for villages, then for cities and states.

The advancements in operations outlined above enabled higher population densities and population growth. The global population of less than 5 million hunter-gatherers has been estimated to have quadrupled in the 5,000 years following the start of the agricultural revolution 11,500 years ago [301]. Another 5,000 years later, it has been estimated to have exceeded 200 million, and by the time of the industrial revolution, some 300 years before present, the global population had reached almost 600 million [301].

Even without specific data on (per capita) GDPs in ancient times, we can readily conclude from the population growth numbers that the global GDP increased by at least two orders of magnitude between the agricultural revolution and the first industrial revolution. What is more, improvements in agricultural human labor productivity caused the relative number of population members working in the primary agricultural sector to decrease and the relative population adding economic value in other sectors to increase. Per capita GDP thus must have increased as well from the days of the first farmers to the onset of the industrial revolution. Whether this global economic growth translated into economic growth for the poor is discussed in the next subsection on social inclusion.

The agricultural revolution appears to have been indismissible for this economic development. It is hard to envision hunting and gathering operating models that can facilitate comparable economic growth, for instance, because of the low population density required for effective hunting and gathering operations. We may recall that many large animal species had already been hunted to extinction at the onset of the agricultural revolution, a development that may have co-triggered the agricultural revolution.

It is also worthwhile to look forward and take a snapshot of the contribution made by the agricultural sector to the global GDP at present. By 2019, the agricultural sector employed slightly less than one in four of the global workforce and contributed 3.5 percent to the gross value created in the global economy for a population of almost 8 billion [38, 399]. On the same thread, we note that in 2018, less than three centuries after the industrial revolution, the metal mining sector created around 1 percent of global GDP [475].

5.6.2 Social Inclusion

The agricultural operations hugely impacted society. It brought a departure from the egalitarian and shared property norms of hunter-gatherers that has not been reversed since. Agricultural societies developed individual and family-based models of ownership, which created inequalities in wealth and power. Economic growth went hand in hand with diminishing social inclusion. Thus, the sustainability effects were mixed.

It has already been mentioned above that the newly emerging ownership models facilitated the accumulation of capital, in particular production capital, to be deployed in operations in the form of land, seeds, animals, tools, and other assets. Egyptian pharaos took personal property and wealth to extremes when having pyramids built for themselves of sizes and heights far beyond the operational capabilities of many later civilizations. Some of the interior rooms of these pyramids were lavishly decorated with gold and other precious metals and stones. Prosperous Phoenician merchant families owned facilities across the Mediterranean and fleets of many vessels. The wealth of these elites was accumulated through effective operations management, utilizing metals, stones, ships, tools, armor, domesticated animals, and human resources.

Human resources could provide labor in return for wages, but also through forced labor and as slaves. Slavery was already common in the cities of Mesopotamia, where people could become enslaved as prisoners of war, as a punishment for crimes, or as a pledge on a debt [584]. Phoenicians importantly depended on slaves for their enterprises. They were involved in raiding slaves, trading slaves, transporting slaves, and deploying slaves in their manufacturing, mining, and household operations [471]. Mining operations yielded particularly harsh and unhealthy circumstances and were mostly performed by slaves. These practices were continued by the Romans, and the cruel conditions were noted by Roman historians who reported the miners to "*long for death as more desirable than life*," thus characterizing an extreme form of social exclusion [240, 471].

Many of the operating models that developed since the onset of agriculture relied on the unequal exploitation of human labor. The city of Tyre hosted 30,000 slaves, forming the majority of the population, when besieged by Alexander the Great [471]. Some evidence suggests that the millions of slaves in the ancient Greek and Roman civilizations outnumbered the civilians [507, 552]. There were many children among these slaves.

Slavery was not exclusive to Mesopotamian city states and the emanating European civilizations. Slavery existed in Africa and developed independently in other societies, such as China and pre-Columbian America [61]. Slavery is one of many forms of coercive labor and, more generally, of the exploitation of humans by other humans as resources for the operations of work and life. Moreover, slavery is still commonly associated with ethnicity, and indeed, the word slavery has racial origins. Slavery most strongly contradicts social inclusion.

5.6.3 Environmental Protection

To construct their agricultural niches, farmers have expanded the practice of late hunter-gatherers to manage land and, in particular, to use fire to deforest land for agricultural operations. It has been estimated that land used for crops and grazing each occupied less than 1 percent of the global land area around the onset of the agricultural revolution. These estimates rose to 2.2 and 5.1 percent, respectively, around the start of the industrial revolution 300 years before present. At present, while the global population is approaching 8 billion, around 12.2 percent of global land is used for crop land and 24.9 percent for grazing, or 37.1 percent of the land on planet Earth. Most of these gains in agricultural land are the result of deforestation.

Agriculture, and especially the increasingly intense farming for the production of animal meat and diary products, contributes significantly to CO_2 emissions and is the main source of N_2O and CH_4 emissions globally. Such in contrast to the forests, which reduce and store GHGs [90]. Thus, deforestation for the purpose of meat or diary production has a double negative impact on net anthropogenic emissions.

For illustration, let us consider early deforestation as a consequence of human operations in Spain. Around 5,000 years before present, the use of fire by early farmers caused a loss of biodiversity, particularly pine forests [569]. Later, the settlements founded by the Phoenicians in southern Spain in search of metals brought further deforestation because of the use of wood for metal smelting operations [569]. Ships were the main carriers in the long-distance supply chains of the Mediterranean civilizations. The Phoenicians, and later the Greeks and the Romans, further deforestated Spain when exploiting pine wood for ship building [569]. Meanwhile, mining continued, and agriculture would grow in later centuries as part of the Islamic agricultural revolution and during subsequent population growth.

The example of Spain illustrates global developments of migration, population growth, and further agricultural expansion, which have continued until to date. As outlined in later chapters, agricultural operations often are not sustainable, and current agricultural operations are not moving in the right direction. They contribute to the transgression of a safe operating space regarding biodiversity, biochemical flows, and GHG emissions [518, 553].

The bees that we have considered in Chapter 3 were among the earliest domesticated animals. The earliest evidence of beekeeping is from Egypt and dates back 4,500 years [315]. It has been independently invented in various other locations, and it has been disseminated across Africa and the Eurasian continent, where the honey bee occurred [315]. Domesticated bees can produce honey as well as wax, which played a role in the manufacturing of pottery and metals [315]. Most importantly, perhaps, bees play an indispensable role in the pollination of agricultural plant species, as, for instance, illustrated by the present-day hiring of beekeeper-owned colonies for pollination services [519].

Beekeeping and the honey bee species Apis mellifera were introduced to the Americas around 400 years ago by European immigrants and elsewhere in later stages [519]. The introduction of honey bees by humans to new continents and ecosystems is a form of ecosystem engineering that impacted these ecosystems in several ways and likely reduced survival odds for indigenous bee populations. Beekeeping can also be unsustainable when practiced in ways that endanger current and future generations of the bee populations involved [519].

5.7 Operations Management Perspectives

The transition toward agriculture can be viewed as an adaptive threshold in operations management. Where hunter-gatherers lived in extant ecosystems, first farmers intentionally replaced many of the main plant and animal species living in their created agricultural niches, even through diverting the flows of water by irrigation systems, i.e., through changing the environment of non-living elements within the ecosystem. The effective management of agricultural operations in turn enabled advances in the management of other operations that subsequently emerged, such as mining, pottery, and trade.

It may be impossible to highlight all or most of the advances in operations management that happened between the onset of the agricultural revolution and the subsequent, much more recent, industrial revolutions. Within the agricultural domain alone, the advancements in agricultural operations beyond those made in the defining early millennia of the agricultural revolution have, on some occasions, been such that they are considered agricultural revolutions in their own right. Among the commonly acknowledged agricultural revolutions are the Islamic agricultural revolution, the British agricultural revolution(s), and the American agricultural revolution(s) [371, 426, 594]. For instance, in Great Britain, total agricultural production is estimated to have doubled between 1600 and 1800, while the agricultural work force remained fairly constant [8]. A fourth agricultural revolution is considered to take place at present and will be considered in Chapters 10 and 11 [490].

Below, we highlight two important operations management developments that are rooted in the agricultural revolution and have impacted operations beyond agriculture. We first consider property, and more specifically, the ownership of resources and capital goods. The second development regards the development of facility layout designs that resulted in economies of scale. Both of these developments build on the principle of division of labor, as already introduced in the previous two chapters.

5.7.1 Ownership, Capital, Civilization...and Warfare

Humans living before the agricultural revolution, i.e., hunter-gatherers, understood and respected individual possession, as did other primates and many other species [80]. Basic notions of possession are typically rooted in physical proximity and control at a certain moment in time, e.g., in the case of food in hand when eating, as befits immediate return operations. At the same time, hunter-gatherers likely tended to share foraging returns as a group and to view property as joint, as is still common in extant hunter-gatherer populations.

Sharing with other community members and other collaborative practices appear to have persisted in early farming communities adopting a mixed operating model combining rain-fed agriculture with hunting [206]. Such communities continued to benefit from the sharing of jointly hunted game in a dynamic habitat, which promoted effectiveness and reduced the risk of these direct return operations.

In other ecosystems, early farmers developed irrigation-based agricultural operations and combined them with livestock farming, as their ecosystems yielded poor hunting returns [206]. These operating models benefited especially from operations management practices such as the coordination of irrigation, which eventually resulted in production surpluses. Farmers operating in such irrigated ecosystems, with their growing, settled populations and higher population densities, developed novel ownership concepts alongside their delayed return operating model. They developed values and norms of ownership, ascertaining that investment in resources employed, such as land, irrigation works, tools, seeds, et cetera, yielded a worthwhile delayed return in the form of ownership of the food produced, such as inventories of grains, vegetables, and animal produce [72, 80].

Thus, we see how the agricultural revolution, and more specifically, fully fledged farming, introduced the ownership of assets and capital goods. Operations management thus started to include the management of capital, of land, animals, machines, and tools owned and deployed in operations. Being in control of the operations, i.e., managing the operations, enabled the owner who made the investment to obtain the intended returns. Personal ownership thus started to form an important development in operations management, which had a self-reinforcing effect. From the time of the first fully farming-based societies onward, humans would not only work and live in constructed niches, but some of these humans would consider parts of these niches to be their property and manage them accordingly.

The quote opening this chapter from the Law Code of Hammurabi illustrates how values and norms of ownership became encoded in law in Babylon, where Hammurabi ruled. Inhabitants of Babylon and other cities developed and adopted a variety of operations other than agriculture and mining and managed these operations, practicing division of labor, specialization, and economies of scale. Some cities, communities, families, and individuals were more successful at newly developing operations management than others and accumulated more wealth. The resulting inequalities necessitated the protection of property against various forms of expropriation, ranging from theft to warfare. Conversely, one might view the concept of private property as making operating models based on theft, violence, and even warfare potentially more rewarding. To ensure the sustainability of prosperity, the early civilizations institutionalized norms, developed regulations, and practiced public service operations such as law enforcement, military services, and public infrastructure services (such as reliable public irrigation systems, city walls, and tax collection [1, 568]).

Evidence strongly suggests that war, i.e., lethal conflict among organized, armed and opposed social groups, was rare among hunter-gatherers and only increased in prevalence and intensity after the onset of the agricultural revolution [115, 196]. A variety of contextual factors associated with the agricultural revolution, such as sedentism, increased population density, accumulation of goods and wealth, social differences, and hierarchy, contributed to the likelihood of war [196].

The practice of warfare independently developed in Mesopotamia (5,000 years before present), China (4,500 years before present), and Mesoamerica (3,000 years before present) [115]. Warfare developed into an operational discipline of its own with its tools and operations management practices. Metal weapons and armor importantly influenced military operations as they became more advanced and voluminous. The battle of Meggiddo, 3,500 years before present, likely involved more than 10,000 armed forces [341]. It evidences the advanced military operations management capabilities of the Egyptians, including the planning and execution of large-scale and long-distance military expeditions with the corresponding lines of command, inventories of weapons, food supply chains, et cetera. The operations management advances of the agricultural revolution brought economic and social advancement, wealth, and effective large-scale warfare.

5.7.2 Operations Management Foundations

Hunter-gatherers operated mostly in extant ecosystems with limited abilities for significant niche construction. Pars pro toto, this held through for the industrial sites in which they produced their stone tools and the locations in which they later produced more advanced instruments made of a variety of raw materials. Without the limitations of a non-sedentary lifestyle, the first farmers constructed irrigated farm lands as well as settlements with buildings. These buildings included communal places, which also served to store inventory. Within the houses the first farmers built for themselves, they dedicated space for the manufacturing of clothes and pottery. Household pottery workshops with multiple dedicated rooms developed over time in cases of sufficient demand for a family to specialize [167].

The operations facility design thus interacted with early organized forms of labor division. Scaling up the specialized operations, these industrious workshops of the early agricultural civilizations would develop into larger facilities [167, 254]. The layouts of these facilities are presently categorized as **process layouts**, as the layout logic dedicates space to machines and tools for a common process type. This applies, for instance, to pottery rooms in which pottery wheels and supporting tools facilitated specialized, highly skilled human craftsmen and women to operate. This type of process layout has become known as a job shop, as positioned in the product process matrix of Figure 5.3. The job shops, for instance, facilitated the scaling up of production from the very low volumes of products for household needs to modest volumes of more standardized products at the village level.

The earliest evidence of "*mass production*" of pottery for distribution along the Mesopotamian trading networks is from more than 5,000 years before present [414]. It is not clear whether these larger-volume operations followed the aforementioned job shop layout—as would presently be viewed as inefficient—or whether

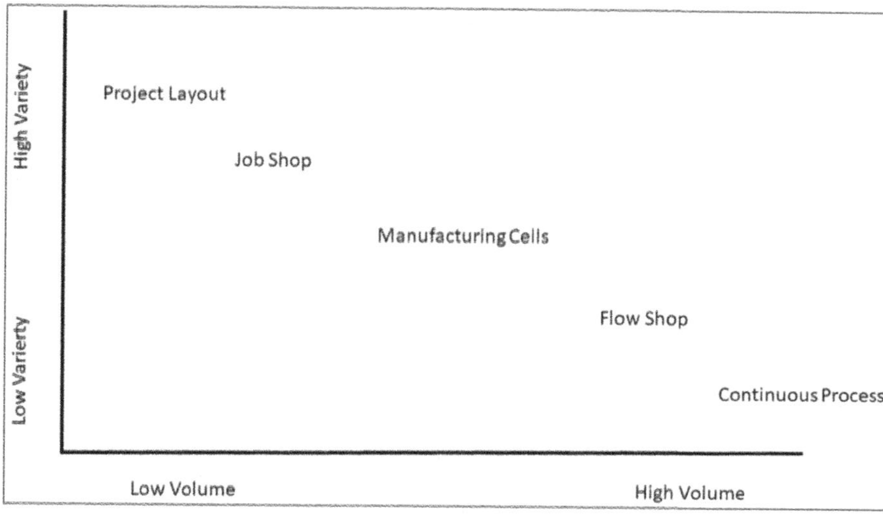

Figure 5.3. The product process matrix that relates contemporary production layouts with production volumes and varieties (based on [510]).

a flow shop layout was already adopted. In a flow shop layout, the resources are organized around the flow of operations for a collection of similar products (rather than around a common process type). Flow shop layouts will be discussed extensively in Chapters 7 and 8. Flow shops typically involve further division of labor and lower-skilled workers to repetitively perform a limited sequence of standardized tasks.

The metal manufacturing innovations of the Bronze Age involved carefully engineered resources such as furnaces. Moreover, these operations followed a complex series of process steps and clear operating standards with narrow margins of error [482, 483]. The metal workers able to operate these processes held advanced, specialized knowledge. While these workers naturally developed operational capabilities through experimentation and experience, extensive instructions to next generations and to other communities must have been provided to pass on the operational knowledge [483]. As the division of labor advanced, extensive training became increasingly important for knowledge and skill acquisition.

The farmer communities dedicated community members to learning and performing these metal manufacturing operations. This is especially noteworthy, as initial applications mostly regarded spiritual purposes and adornment and were of little use for subsistence. Over time, a widespread mining industry arose, which continues to be a vital base industry until today.

Advances in operations management also accompanied the realization of large-scale, unique public infrastructural projects. Well-known examples include the city walls and tower of Jericho (the first walls date back to 10,000 years before present), the irrigation works along the Euphrates and the Tigris (from 5,000 years before present onward), and the construction of the pyramids in Egypt (from 4,600 years before present onward) [39, 172, 491].

While some of these infrastructural projects started on a small scale, their importance and size increased over time [172, 491]. The building of pyramids especially stood out for its magnitude and complexity. It has been estimated that the construction of the larger pyramids would require 20 years and up to 10,000 workers [145, 172]. The Great Pyramid, begun by King Khufu, was almost 150 meters high, and the base occupied 230 square meters. It was built of more than 2 million stones of more than 2 tons each [172]. The builders left no written documentation on the highly complex and extensive construction operations of the pyramids.

Of all these large infrastructural projects, especially the pyramid construction projects, subsequent civilizations, including the ancient Greeks and Romans and the present global population, are wondering about their engineering, project planning, resource planning, and execution, in other words, project operations management [145, 168, 172]. In any case, they form evidence that advanced project layouts (see Figure 5.3), in which all operations are organized at the location of the project (e.g., pyramid construction), emerged in agricultural societies, in addition to job shop layouts and possibly flow shop layouts.

Especially the operations of such large projects and of mass-produced goods benefited from non-inclusive operating models in which some members of the human population functioned as human resources in the operations managed and owned by others. Operations management even involved the management of owned human resources, such as slaves. We will see in subsequent chapters that such non-inclusive, inequitable operating models and the management of these models have continued to play a significant role along the timeline of operations.

DOI: 10.1561/9781638282853.ch6

Chapter 6

The Production of Visible Language

The first use of writing,..., was a means of control upon the delivery of goods and ultimately a control on the production of real goods.

Denise Schmandt-Besserat [509]

6.1 Introduction

In Chapter 4, we have already discussed how brain size, cognitive abilities, and other physiological characteristics of hominins are co-evalled with advances in foraging operations. The development of speech and—closely related—language has been an important element of this process. There is little direct evidence of the evolution of speech and language, and it is not known whether speech is exclusively practiced by humankind (or whether, for instance, Neanderthalers also communicated using a spoken language). It is widely believed that humans communicated through speech 50,000 years before present [270].

While gestation, vocal sounds, and other forms of communication are commonly found among animal species, the ability to use spoken language is considered to distinguish humankind from other extant species. Spoken language has developed in close alignment with human operations and operations management as it has enabled collaboration, coordination, instruction, problem solving, et cetera.

We now address a collection of further advancements, namely the production of visible language. The most common form of visible language is the script. We shall see below that precursors to the script were developed for operations management purposes shortly after the agricultural revolution. This chapter covers further advancements in the production of visible language until the first industrial revolution of the 18th century.

6.2 Development of the Script

Some of the early agricultural civilizations developed a token-based system to record information around 10,000 years before present, which has led to the understanding that "*the need for recording was related to particular aspects of human adaptation to food production*" [509]. It has been argued, however, that storage of agricultural goods as such, trade of agricultural produce, and accumulation of wealth necessitated the use of tokens, as these developments advanced elsewhere in several early agricultural societies without tokens. Rather, the use of tokens appears to have developed in societies of all sizes that required a central authority that redistributed food beyond direct ties and coordinated and kept records of the contributions of foods and other products for shared ritual purposes [179]. Such tokens, as depicted in Figure 6.1, were used to denote quantities and used numerical systems to count objects and measure capacities. This system may have developed toward engraving the same visuals on clay balls, which were sealed and enclosed in clay envelops [179], as illustrated in Figure 6.2.

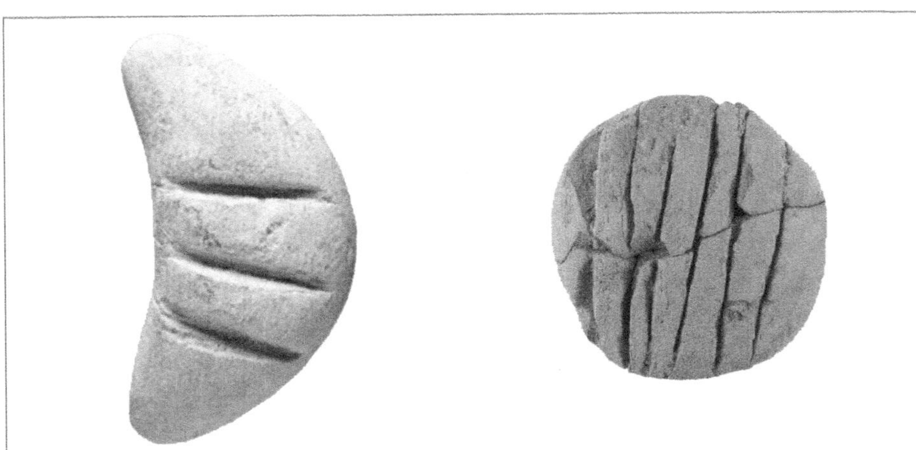

Figure 6.1. Tokens for wool and silver with numerical engravings—Courtesy of the Institute for the Study of Ancient Cultures of the University of Chicago.

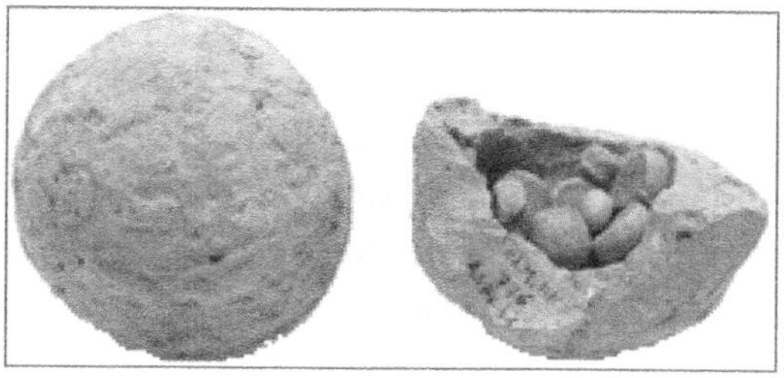

Figure 6.2. A clay envelop in the form of a ball with its content of an engraved token, 5,100–5,300 before present, Chogha Mish, Iran—Courtesy of the Institute for the Study of Ancient Cultures of the University of Chicago.

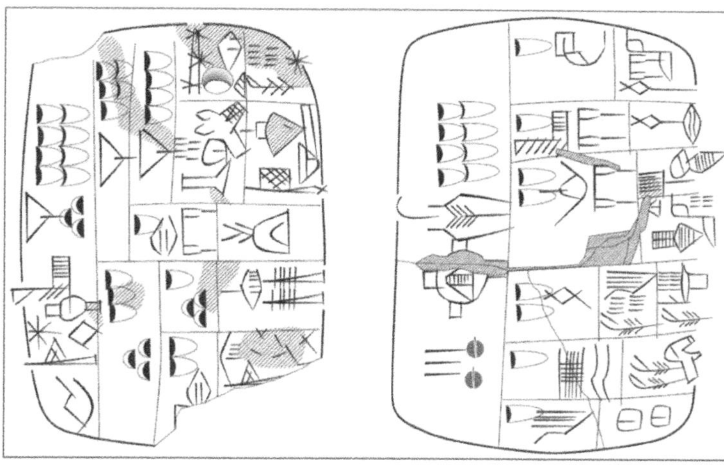

Figure 6.3. Two clay tables with cuneiforms, each of which specifies a "herd" of eight slaves as human resources (reproduced from [179]).

The transition from inscriptions to clay tablets that started 5,500 years ago is viewed as the first script in which the symbols are called protocuneiforms [179, 619]. The oldest of these clay tablets describes inventory records and trade transactions. They contain basic information on quantities of objects (goods and land), actors involved, and the professions and locations of these actors [179, 619]. Goods and animals were often referred to in curvy pictographs.

The script first developed in the Mesopotamian city of Uruk, which is considered to have been the world's oldest city, reaching a population of 20,000 to 50,000 on 2.5 square kilometers. Within a few centuries, this script developed into a straightened set of symbols, called cuneiforms, that were written on the clay tablets through the use of standardized reeds [619]. Cuneiforms supported operations management, enabling the keeping of inventory of goods produced, traded, and bought and of the required resources. For instance, Figure 6.3 depicts the administration of two “herds” of eight slaves [179]. The use of script for religious texts, letters, and literature would also follow within several centuries [619].

Uruk’s civilization, however, turned out to be unsustainable, and the city collapsed not long after (and despite of) developing the script [619]. The script survived and would form the basis for other Near Eastern scripts and eventually form the basis for European scripts. Writing developed more or less concurrently and independently in Egypt and, in later stages, in China and Mesoamerica. The writing systems in China and Mesoamerica initially developed mostly for religious purposes rather than for purposes of inventory and transaction record-keeping [619].

6.3 Writing Operations and Resources

The creation and operation of these first information systems themselves are worthy of consideration as such. These information systems required operations to produce the physical clay tablets, the inscriptions in the tablets with standardized tools, and the use of standardized cuneiforms, sealing, and subsequently hardening for future reference. Alternative information carriers that have been used for cuneiforms are stone (for instance, for the code of Hammurabi), wax, and metal [1, 619].

The concurrent visible language production developments in Egypt made use of the innovative product papyrus as the main material to carry information [211]. The Egyptians manufactured papyrus from river plants widely available in the Nile but less so in other Mediterranean areas. The limited accessibility and durability of papyrus led other civilizations to prefer parchment, which is manufactured from sheepskin [211].

While the information carrier clay was produced in the form of tablets, papyrus and parchment were initially produced in scrolls. Compared to clay, the technology of scrolls brought many operational advantages in production, storage, and use, especially for longer texts. As civilizations advanced, longer texts became of increasing importance to record and disseminate information for legislative purposes, religious purposes, personal communication, science, and literature.

Around 2,000 years before present, the codex was invented as an alternative to the scroll. Codices were the direct predecessors of books, consisting of pages with text on both sides, bound on the left side, and initially with a wooden cover [256]. The uptake and spread of codices was particularly common among early Christians, and codices more generally replaced scrolls by the turn of the 4th century [256].

Around the same time, postal services emerged in the Roman Empire to spread announcements and decrees. The required manual labor of copying texts was mostly performed by slaves in the Roman administration [211]. In early Christianity, the manual copying of religious books was a religious art work to be performed by monks. Working from their scriptorums in monasteries, monks would become the main human resources involved in book production operations for many centuries to come [211].

6.4 Paper and Printing

Paper was invented in China slightly less than 2,000 years before present. As the Chinese chose not to disclose the manufacturing process, it would take many centuries before the use of paper reached other Asian civilizations, and perhaps almost a millennium before paper manufacturing reached other continents such as Europe. The knowledge about the development of the script in the Americas is scarce, as the materials of some native American writing systems have been destroyed while others still remain to be deciphered [211].

The Chinese government implemented a postal services system in the 10th century, not long after xylography, the oldest known form of printing, had been invented in China [211]. Xylography is a mechanical process by which a full page is printed at once using a carved wooden block and ink. The invention of xylography spread through Asia, reached the Muslim world, and eventually spread to Europe via Venice [160].

It is unclear whether Johannes Gutenberg had knowledge about this technology when (re)inventing book printing in the 15th century. Most likely, Gutenberg was one of several craftsmen who built on already existing (partial) printing techniques when developing his famous printing press [193]. The details of his original press have been lost, yet an imagined replica of this complex device is depicted in Figure 6.4.

In comparison to xylography, the increased complexity of the printing mechanics developed by Johannes Gutenberg enhanced the flexibility and ease of application of the printing process. It required the development of letter stamps to produce matrices for the letters and to arrange these letters for printing with ink. A closely aligned set of advancements in metallurgy, printing (pressing) mechanics, and ink were needed for printing, and it may well be that some of these advancements were completed after Gutenberg's initially developed set up [21].

The human-powered printing press has been viewed as industrializing the monastic book-writing craft. Its capacity for (high-volume) mass production positions it as a precursor to the industrial revolutions that would emerge three centuries later. The invention of the printing press has also been classified as a revolution of its own. It has been classified as a printing revolution, a book revolution, an information revolution, a communication revolution, and a knowledge revolution [177].

Figure 6.4. Replica of the (lost) original printing press (source [394]).

Printing yielded great efficiency gains in the production of books. Already in the 16th century, a printer could produce around 3,000 (one-sided) pages per working day (of 12 to 16 hours) [193]. Together with the increased production of paper, which was lower in cost than parchment, these efficiency gains brought substantial operating cost reductions, which translated into a lower cost of the books sold and lower prices, increasing the demand for books. Aided by a rapid spread of print shops throughout Europe, the operational efficiency gains resulted in a dramatic increase in the volume of books supplied [160]. The average number of books produced per year in Europe increased from less than 30,000 in the 14th century to more than 2 million in the 16th century [86]. As a direct effect of these developments, new industries emerged, such as the paper industry and the printing industry, with professions such as printer, writer, bookshop keeper, et cetera. Meanwhile, jobs related to (re)production of books by hand became obsolete.

6.5 Sustainability of the Script and Book Printing

6.5.1 Economic Development

It is not easy to assess the sustainability effects of the development of the script and book printing from the first protocuneiform use onward. There is no counterfactual to economic development, social inclusion, and environmental protection on a planet without scripts and information carriers. On the other hand, it is hard to envision the civilizations that have adopted tablets, scrolls, books, et cetera, operating without these information carriers. Could the initial trade routes have developed into larger-scale networks without inventory records? Could the property concepts underlying the deployment of capital goods and trade in finished products have been implemented and maintained without transaction record-keeping? These information carriers facilitated not only the economic development of civilizations but also the institutionalization of trade between these civilizations, for public and private actors alike.

Among these information carriers, books stand out for their contribution to the dissemination of knowledge, which forms a means of disseminating novel operational practices and subsequent further innovations. The impact of the advancements of Gutenberg and his successors reaches far beyond efficiency improvements in book production and increases in sales volume. The printed books greatly improved the availability and accessibility of production factor knowledge. Reduced book prices have positively influenced education enrolment in the UK from the 15th century onward. Education and knowledge have, in turn, positively contributed to economic growth [358, 359]. Using a global sample of countries, Baten and Van Zanden [46] find that "*books per capita had indeed a strong, positive, and economically significant impact on welfare growth*" for the time period 1450–1849. This causal relationship between the number of books printed and economic growth also has significance for large parts of the Muslim world, where regulation by the Ottoman rulers severely limited adoption of the printing press until 1729 [129].

6.5.2 Social Inclusion

Book printing made access to knowledge more inclusive because it enabled the mass production of books at low prices for mass audiences, including the lower classes [358]. However, literacy was limited across the globe in the 15th century when the printing press emerged and remained so for many centuries to come. Hence, literacy and access to books may have widened divides in welfare and well-being between the literate and illiterate subpopulations, both within countries and among countries [46, 129]. Likewise, one may view that the ancient public administration and regulation of trade and inventory transactions and property of resources, among which slaves, promoted inequitable, non-inclusive operating models. The successful administration of these business models may have generated prosperity for some yet inhibited economic growth for others with little access to ownership of resources or being considered an owned resource. Thus, the early advancements in information technologies appear to have had mixed effects on social inclusion.

6.5.3 Environmental Protection

While initial increases in the use of paper may have been modest enough to be absorbed by natural forest growth and forestation operations, the paper and pulp industry has over time increasingly practiced unsustainable production operations. It contributes to CO_2 emissions, deforestation, the use of scarce water resources, and chemical waste [436]. The paper and pulp industry consumed 5.7 percent of global industrial energy use in 2004 [173]. Moreover, over the full life cycle, paper produced 1.3 percent of the global GHG emissions in 2012 [186]. The paper and pulp industry is not on track for net zero emissions by 2050 [2].

Not all of these environmental sustainability challenges can be attributed to the production of visible language. More than half of the paper is produced for other purposes, such as packaging [173]. Chapter 9 sheds light on the net anthropogenic emissions gains that can be obtained by switching from paper printed-books to providing digital access (either as a product or as a service).

6.6 Operations Management Perspectives

The relevance of the developments described in this chapter from an operations management perspective is two-fold. First, there is the sequence of advancements in the designs of the information products and the operations to produce them. While they have gradually developed over many thousands of years, some of them were imaginative and highly influential innovations. This applies, for instance, to the use of clay tokens and clay envelops for keeping inventory and transaction records in early agricultural communities.

The highly innovative practice of keeping inventory and trade records enabled a second relevant development through its positive impact on the effectiveness of the operations of the agricultural revolution and beyond. Inventory and transaction recording have remained a core practice in business operations globally ever since (see also Chapter 9). It promotes division of labor, productivity, trade, and so on. The fact that the symbols and record-keeping practices would develop into the first script in Mesopotamia further evidences the core role operations and operations management have in society.

While this chapter illustrates that book printing developed gradually and over many continents and centuries, the printer developed by Gutenberg was a breakthrough in operations in its own right. It is a highly complex machine by which humans can mechanically perform a printing process, a sequence of operations that is quite different from the way human resources copy and write books, whether slaves, monks, or otherwise. While fully operated by humans, and hence with human energy resources, it industrialized the book production process.

Before the introduction of xylopgraphy and the printing press, book production was a project, positioning it in the upper left-hand corner of the product process matrix in Figure 5.3 [511]. On occasion, handwritten books would be written in low-volume batches over a time horizon of years, with a very basic job shop layout in which the person writing formed the main resource. With xylopgraphy and printing, the press became the central resource. Subsequent steps to produce books, such as binding, would be conducted afterward with a different set of resources. The thus-arising flow shop layouts facilitated the production of books in larger batches and allowed books to be mass produced. The printing press moved the production of visible language down the diagonal of the product process matrix, which is a key development in the life cycle of the production processes for many goods and services [511].

It has already been mentioned that, however ingenious, the main relevance of the printing press for operations is beyond book production. It served as the basis for a faster and more widespread exchange of knowledge and hence increased the availability of production factor knowledge, which in turn promoted economic development. Books also improved access to operations management knowledge, e.g., regarding inventory management, new process techniques, new process designs, et cetera. More profoundly, the printed books contributed to advancing science, and specifically the sciences that have laid the foundations for the industrial revolutions in operations covered in Chapters 7 and 8. Reading this book, even if in a digital format, as is possible since the 3rd industrial revolution, may show that books still serve as a valuable resource to exchange knowledge, including knowledge about sustainable operations management.

DOI: 10.1561/9781638282853.ch7

Chapter 7

The Global Cotton Supply Chain

The Cotton Mill

Hum, whirl, click, click, clatter
Rolling, rumbling, moving matter
Whizzing, hissing, hitting, missing
Pushing, pulling, turning, twisting

Buzz, bang, going, coming
Standing, creeping, walking, running
Piecing, breaking, starting, stopping
Picking, mixing, fixing, copping

Push, rush, cleaning, oiling
Slipping, sweating, screaming, toiling
Fetching, taking, spoiling, making
Saucing, swearing, bagging, bating

Here, there, this way, that way
Bad-end, nar-here, fur-on, up-there
Break-it-out, wind-it-off, hurry piece-up
Get-em-up, quick, or a'st ha' to stop

Steam, dust, flyings choking
Stripping, grinding, brushing, joking
Full time, short time, no time – so that
Enough's in a mill without Surat!

N.N., September 24, 1864, The Bolton Chronicle, Bolton, UK [401]

7.1 Introduction

A valuable perspective on the evolution of operations is to view it as a journey in energy efficiency [626]. The control of fire already directly exemplifies this perspective. The process of cooking food with the energy of fire in the form of heat reduces the human energy needed for the digestion process. Hence, the control of fire enabled hominins to consume more energy and to use relatively less energy for digestion, resulting in a net energy gain. The energy gained meant more time for other operations such as hunting, gathering, tool manufacturing, and subsequent innovations. Compared to hunting and gathering, agriculture reduced the effort required to obtain food, again freeing energy for other activities. These energetic efficiency gains took time to achieve, yet eventually proved advantageous enough for agriculture to be adopted in all regions where it was feasible.

Book printing greatly reduced the human energy needed to produce a book. A variety of other mechanical innovations reducing human energy needs for operations happened earlier or around the same time. The Near Eastern medieval Saqiya is a waterpump driven by an ox [265]. Water mills were commonly operated in medieval times for the grinding of grains and corn, crushing ore, and paper production. In ecosystems without running water, windmills have been operating since the 7th century [265]. These and other machines were documented in the 13th century by Ibn Al-Razzaz Al-Jazari in The Book of Knowledge of Ingenious Mechanical Devices [266]. It can be viewed as a seminal text on mechanical ecosystem engineering.

Water and wind are free and renewable energy resources that can power operations. The renewable energy resource wind has long been the main resource to facilitate long-range travel by boat. The Phoenicians already made extensive use of sailing vessels and reached destinations as far as Senegal and Ireland [92]. The Spanish and Portuguese took initiatives to let the wind sail them to India by the end of the 15th century, as they viewed the land routes, the Silk Roads, as inefficient. A sea route to India was believed to improve access to India's valuable goods, in particular spices [413, 607].

In one such initiative, Columbus and his men bravely sailed out to reach India via a Western route, which he believed to be the shortest. Columbus failed to achieve his original objectives and landed on a Caribbean island instead. As is well known, his endeavors would, however, reestablish connections between the Eurasian and American continents and societies. A couple of years later, Vasco da Gama successfully sailed out to reach India via an east-bound route along the African shore. It would take another hundred years until Magelhaes was the first to find a Western route to India, around the Americas.

The newly established sea routes were quickly exploited and expanded. Spain and Portugal established new trading routes and colonies, bringing in valuable goods from overseas. The Spanish ships returned from Latin America filled with precious metals that were partially used to procure spices in Asia. In addition to supplying spices, the Portuguese started to supply textiles from India [413, 607].

The global supply chains have now quickly shifted to the more efficient naval routes, diminishing the importance of the Silk Roads as supply chains of spices, textiles, and other goods. Other European countries, such as France, Great Britain, and the Netherlands, quickly followed the Spanish and Portuguese examples to sail the oceans. They too started operating fleets and trading posts, and they all set up colonies. Eventually, the increasing volumes of goods flowing in along these new supply lines triggered the industrial revolution.

7.2 The Industrial Revolution

As many as 50 different species of cotton have existed on planet Earth for 5 to 15 million years, among which are four domesticated species [598]. These four species have been independently domesticated and cultivated since more than 5,000 years ago by humans living in (semi-)arid (sub)tropical ecosystems in Africa, America, and the Americas [598, 629]. Cotton became an agricultural resource to produce everyday goods such as ropes and clothes and was already widely traded in Mesopotamia, where the origin of one of the domesticated species lies. Textile manufacturing for clothing dates back around 3,500 years in North America and India [526]. The indigenous humans encountered by Columbus after landing in the Caribbean wore cotton clothes, which confirmed his hypothesis of having reached India and its precious fabrics [629]. In the 3,000 years between the establishment of the cotton industry in India and Colombus' voyage, Indian cotton textiles would be exported further and further west to Mesopotamia, Egypt, Greece, Rome, Spain, and Great Britain, and local production developed in amenable areas [526]. Cotton, however, would play a relatively modest role as a raw material for European clothing until well into the 17th century.

In the naval logistics developments of the 16th century, Great Britain continued to rely on land routes via the Near East for the supply of spices until it felt compelled to initiate sea routes in response to supply disruptions caused by foreign forces. By 1600, they founded the British East India Company to supply spices and other goods from providers east of the Indian Ocean. Soon, the company started to import cotton cloth and raw cotton.

By 1625, the British East India Company had brought in more than 220,000 pieces of cloth [484]. Because of the popularity of cotton, the imports raised concerns among wool and linen manufacturers, who called for measures to protect their business operations. Around 1700, a century after the company had started to operate, imports of cotton (and silk) fabric and cloth were banned with the exception of white cotton (banned in 1721) or for re-export [77, 484].

As these regulations were gradually relaxed around 1750, the cost price of Indian cotton products was less than half of the cost of British products (produced from imported raw cotton) and of superior quality [484]. The nominal value of imports of cotton goods would continue to increase and was almost ten times higher in 1795 than it had been before the protective measures were implemented [77]. By then, the industrial revolution of steam-powered cotton mills covered in detail below had reached a threshold, and British cotton manufacturing operations became competitive with manually produced Indian cotton in quality and cost.

Originally, the operations of producing fabric from raw cotton involved various laborious steps that required manual, skilled labor. After harvesting the cotton, the first activity, called ginning, is to remove seed debris, after which fibers for further processing remain. Next, these fibers are combed into alignment using cards, resulting in rovings. Spinning is then the activity of producing yarn from the rovings using a spindle. The yarn can be woven, possibly with yarn from other materials such as wool or linen, into fabric. Clothes are produced from the fabric.

The coloring can take place on various occasions between these production steps. The first opportunity is to color the yarn and then weave fabric from colored yarns. Alternatively, woven fabrics can be colored, or clothes can even be colored after being manufactured.

Of all these production steps, ginning, spinning, and weaving stood out for their labor intensity. Ginning was a low-skill manual operation in the 18th century and will be covered in more detail below. Spinning and weaving required highly skilled craftsmen and women operating a mechanical device. The first such device, the spindle wheel, had been invented in China and had long formed the standard tool used for cotton spinning in India by the time the European interest in cotton had grown to serious volumes [317]. In the 15th century, it reached Europe, where further advances toward the presently known spinning wheel were made, among others by Leonardo da Vinci [317]. Over the next couple of centuries, a simple human-powered machine developed and became more widely adopted in Europe. These were typically operated by women to produce yarn at home, as contracted by cotton producers and traders [547]. In comparison to the highly efficient and skilled Indian spinners who worked for lower wages, European hand-manufactured cotton fabrics were more expensive and of lower quality [9, 547]. Indian operations have held a strong competitive advantage over European operations so far [9, 526].

As demand for cotton yarn and especially high-quality (fine and strong) cotton yarn rose in England in the 18th century, affordable skilled labor increasingly formed an operational bottleneck in cotton manufacturing. In this context, James Hargreaves, father of a cotton Spinning family, invented the Spinning Jenny in the 1760s [9, 547]. His first Spinning Jenny held eight spindles to spin in parallel and could be operated by a child with limited spinning skills [547]. Soon after, James Hargreaves and others designed and developed larger spinning jennys, to be operated at home and powered by humans.

By 1769, the businessman Richard Arkwright had advanced the mechanization of the spinning process on the basis of different mechanical designs. He developed a machine that—among other innovations—used rolling to draw cotton fiber from the rovings, thus mechanizing a task previously done by skilled hands [9]. He envisioned a scale of operations beyond the capacity of a family home and a scope of operations that encompassed the complete cotton cloth manufacturing process. Arkwright developed a carding machine and developed various layouts and workflows for mass production of cotton [9]. The first layout to be implemented was a horse-powered water frame workshop in Nottingham that started operating in 1772. This workshop used water and animal power to mechanize cotton production, significantly reducing the amount of human energy needed.

In a next step, which was likely inspired by John Lombe's five-floor silk-throwing plant in operation since the 1720s in nearby Derby, Arkwright built a five-story water-powered cotton manufacturing plant in Cromford next to the Bonsall Brook [116, 625]. The layout design for the water-powered plant was complicated by the demands for water and steam to power the machines. The power system of the spinning "mill" (i.e., the plant) used an extensive and space-consuming system of shafts and belts emanating from the central wheel to put each of the machines into motion. The system brought limitations to the locations of the machines. Hence, the layout was a compromise between the requirements of the power system and the optimization of the workflow of the primary cotton production operations. The complexity of this design problem was such that the first designed and constructed factory appeared to be unsuited, and Arkwright quickly resorted to a second factory, or spinning mill.

While Arkwright's first plant never reached the operating phase, the more successful second plant built next to it started operating in 1776. This Cromfort Mill hosted several water frames and a total of more than 1,000 spindles, sometimes running concurrently for 24 hours a day. It would employ a workforce of 450 employees. Many of the employees were women and children, family members of the miners working in the nearby lead mines [625]. Arkwright built houses to host this large workforce in the small village of Cromford. Work was tightly organized, with working days of 13 hours starting at 6 AM (7 AM in winter), working in two shifts. The human resources were now embedded in a cotton manufacturing operating system that was largely powered by water.

Samuel Compton, who had worked on the spinning jenny as a boy, would develop a third important innovation by developing the spinning mule around 1780. The spinning mule was able to operate on a large scale and, different from the water frame, produced high-quality, strong, and fine yarn [9].

By the end of the 18th century, the first steam-powered spinning mules were in operation. This completed the energy transition in cotton production from being fully powered by human resources, via being powered by horses and water, to steam-powered mass production. Soon, the steam-powered mills produced cotton of higher quality and at a lower cost than cotton produced manually in India. The competitive advantage had shifted to England.

To provide a quantitative flavor of the speed of this revolution in the cotton industry, let us mention that in 1784, 347,000 spindles were in operation in Great Britain, 82 percent on (typically home-operated) spinning jennys and 1 percent on mules. Twenty-seven years later, the number of spindles had grown more than tenfold to 4.7 million, with more than 90 percent spinning mules and none spinning jennys [366]. As a result of the mechanization and use of steam power, labor productivity in cotton manufacturing increased by an unprecedented factor of more than 100 within the 70 years from 1760 to 1830 [366].

7.3 Global Cotton Supply Chains

The advancement of the cotton manufacturing operations in Great Britain impacted the operations of supply chain partners and competitors in Europe, Asia, Africa, and the Americas. Let us start considering these developments from a chronological perspective, and hence in India.

India had been exporting cotton fabric and clothes for multiple millennia when the Portuguese established the sea routes to Asia, which opened up opportunities to trade with East Asia other than via the Silk Roads. Cotton arriving in Europe via the Silk Roads had been expensive because of the monopolistic markups of the powers controlling these roads. Hence, the sea routes provided an opportunity to bring cotton to Europe at a much lower cost. On the Indian side, the sea routes thus meant a huge opportunity for the cotton industry, and they brought prosperity and growth to the extent that Bengal cotton spinners were considered to be wealthier than British spinners in the 18th century [484].

This situation changed drastically when the British East India Company established a monopoly on the cotton trade and forced Bengal weavers to supply below-cost cotton by 1770. Soon, however, the industrialization in Great Britain reduced the competitiveness of the cost and quality levels of Indian handmade cotton [484]. Indian imports would fall by half within a decade and continue to decrease over the 19th century. By 1850, the value-adding spinning and weaving operations for which India had stood out for several millennia had disappeared from India's export figures, and the relative contribution of raw cotton to India's exports had quadrupled [484]. The direction of the cotton supply chains had been partially reversed, and the majority of Indian cotton cloth consumption was imported from Great Britain by the end of the 19th century [77].

The British's growth in raw cotton imports procured for the growth in production volumes far exceeded the growth in raw cotton exports from India. The main source of cotton imports for Great Britain shifted to the West, across another ocean. Increasingly, the bulk of the imports came from the Caribbean islands known as the West Indies and from North America.

In 1790, the young nation of the United States of America exported a modest 12,000 pounds of raw cotton to Great Britain. At that time, the ginning of the cotton, i.e., the separation of the lint from the seed, was a time-consuming manual operation for the cotton species (Gossypium barbadense) grown in the USA. Human productivity was in the order of a pound per day [252]. For other cotton species, ginning devices had been successfully developed some 2,000 years before present and applied across the globe, including India, China, and, in a later stage, Europe [324]. These relatively simple, man-powered machines were also being used in the United States on a small scale in ecosystems amenable to these cotton species but failed to be effective for the cotton species Gossypium barbadense.

Among the many initiatives to develop cotton-ginning devices for the species Gossypium barbadense, the 1793 machine design of Eli Whitney and his partner Phineas Miller stood out for its productivity gain [629]. It enabled a single person to produce as much as 50 pounds per day and resolved the problem of cotton ginning being "*tedious and unprofitable*" [252]. Subsequent improvements by the many adopters involved animal and steam-powered cotton gins, which produced up to 1,000 pounds per day [629].

In the 10 years between 1790 and 1800, US cotton exports to the UK grew from 12,000 pounds to almost a million pounds. By 1860, the USA had exported well over a billion pounds of cotton, more than half of the cotton it produced. The cotton export represented more than half of the US export value [32]. From the British

Figure 7.1. Girl working in a cotton mill, South Carolina, USA (source Library of Congress, USA).

perspective, US cotton formed almost 90 percent of all cotton imports [32]. The mass production operations of the cotton supply chain importantly drove the economic advancements of the leading economies on both sides of the Atlantic Ocean. Figure 7.1 shows an early 20th-century girl between the machines she is operating in a US cotton mill, illustrating the advancements in industrial operations.

The above numbers reflect a tremendous growth in US agricultural cotton production, from 1.5 million pounds in 1790 to 2.3 billion pounds in 1860. As is well documented, the human workforce required for the quickly increasing volumes of cotton growing at competitive prices consisted predominantly of African slaves. The total number of slaves grew from 700,000 to 4 million over the same time period [32]. The invention of the cotton gin, which mechanized the cotton ginning operations and intended to reduce the need for costly human labor, sparked a development that effectively grew the workforce demand for cotton growing. It increased rather than reduced slave labor in the cotton supply chain.

The tight linkage between American cotton farming and British cotton manufacturing has been characterized as triangular. The third angle rested in Africa, where the cotton supply chain was integrated with the workforce supply chain from Africa to America. The companies "buying" slaves in (mostly East) Africa commonly paid with cotton clothes and fabrics. Initially, the cheaper and higher-quality Indian products were most attractive for traders. However, as the quality and cost of cotton improved in Great Britain and the Indian cotton industry worsened, volumes of British cotton traded for slaves steadily increased. Thus, the triangular business routes of shippers in these supply chains might depart from a British harbor, bringing cotton to Africa. There, the cotton would be traded for slaves brought to the

Americas. From the Americas, the ship would then return to Great Britain with raw cotton.

The cotton products manufactured from this raw cotton would not only suffice to load a next ship bound for Africa but would also provide cotton for the British market, the European market, and the Indian market. The profits of entrepreneurs involved in these cotton supply chains were such that Richard Arkwright is known not only for his innovations in cotton manufacturing operations but also for becoming the richest entrepreneur of the industrial revolution [9, 511].

7.4 Coal

Cotton was not the only good for which global supply chains expanded in the early decades of the industrial revolution, while connecting agricultural areas on one continent with manufacturing on another and with consumers on these and other continents. Spices, tea, sugar, and linseed are among the other goods for which comparable high-volume mechanized manufacturing processes and supply networks spanning the globe developed to provide large consumer populations with high volumes of products. The machines at the heart of these mass production operations were importantly powered by steam engines rather than by renewable resources like wind and water, or by animals or humans.

While the earliest known reports of experiments with the steam engine date back to the first century and various further advancements have been made over time, the first small-scale steam-powered machines were operated in the mining industry in Great Britain in 1698 [534]. Over the course of the 18th century, British innovators such as John Newcomen and James Watts improved the design of steam engines, resulting in steam engines of significant use in mines (for drainage) and capable of powering manufacturing plants even in the absence of water as an additional energy resource (as was the case for Arkwright's cotton mill). By 1780, steam engines aided the operations of various manufacturing facilities and mines in Great Britain [153, 534].

An important contemporary development was Abraham Darcy's invention of producing iron using coal rather than charcoal in the early 18th century [200]. This invention was particularly welcome as coal was cheaper than charcoal when timber had become increasingly scarce in Great Britain because of deforestation. Together with the spread of steam-powered machines, the increasing demand for iron during the industrial revolution thus stimulated the demand for coal, which soon became the main energy resource driving the industrialization of operations.

The rapidly increasing number of steam-powered manufacturing plants also generated an increase in demand for the transportation of coal. High-pressure steam

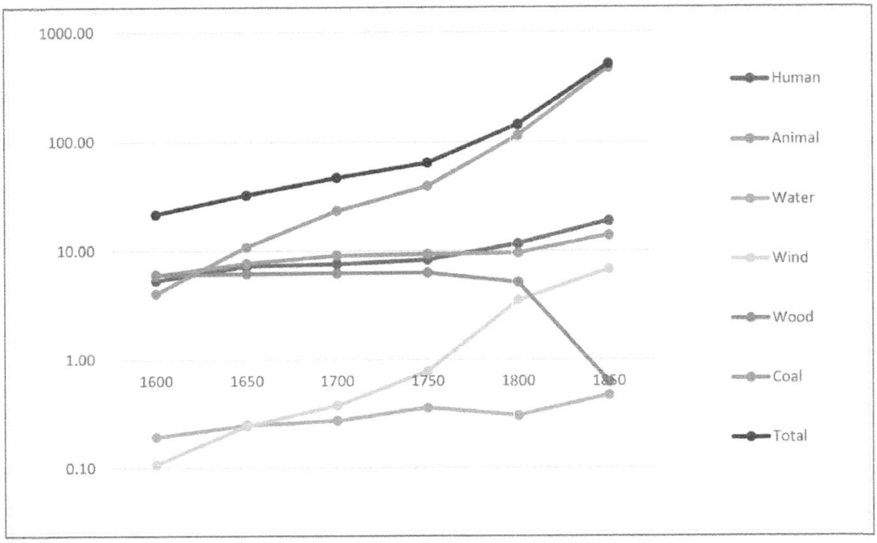

Figure 7.2. Energy use by source in the UK, decade averages, 1650–1850 in TeraWatt Hours (logarithmic scale) (source [626]).

engines small enough to power cars or trains were developed by Richard Threvithick [534]. These engines further improved the applicability and operating cost of coal-fueled steam engines [9]. The use of steam-powered locomotives to pull trains of wagons filled with coal along tracks became an attractive substitute for horse-powered trains and would soon be adopted in the coal mine industry (and elsewhere, e.g., in iron ore mining and for public transport) [534].

Altogether, the application of British coal as an energy resource in mines, transportation, and manufacturing, both domestically and abroad, resulted in an increase in the volumes (produced and) consumed, from around 4 million tons in 1750 to almost 60 million tons by 1850 [451]. It marked the first transition toward the large-scale application of nonrenewable energy resources, as witnessed by Figure 7.2.

7.5 Sustainability of the First Industrial Revolution

7.5.1 Economic Development

The industrial revolution brought Great Britain considerable economic growth in the 18th and 19th centuries. The productivity gains brought by the newly developed steam-powered high-volume operations contributed to a doubling of per capita GDP [76, 278]. As the population almost tripled over the same period, GDP grew more than six-fold [78]. The productivity growth in industry was more than three times the productivity growth in agriculture, which experienced a revolution of its own [78]. No other large European countries realized similar growth over the period 1700–1870. By then, Great Britain had realized the highest per capita GDP in Europe. The British life expectancy and human development index (HDI) scores were also among the highest in Europe by 1870 [430, 446].

The United States of America, established in 1776, almost doubled per capita real GDP from the start of the 19th century until 1860 [373]. As it witnessed a larger population growth than Great Britain, its real GDP growth also exceeded the British real GDP growth.

In India, the per capita GDP was below the British per capita GDP in 1700 and had decreased by approximately 20 percent by 1870 [76]. Thus, while the industrial revolution contributed to economic growth for many of the poor in the UK and the US, it had adverse effects on the economic development of other poor populations, such as the Indian poor, whose poverty was more severe at the onset of the industrial revolution.

In 1820, differences between countries formed a modest component of global income inequalities [103, 382]. These between-country differences would steadily grow, with per capita GDPs doubling in the early industrializing countries within 50 years before 1870. Thus, the per capita GDP increases and population growth in industrializing, "developing," countries indeed increasingly determined and enlarged global inequalities, as confirmed using a variety of data sets and measures [103, 123, 382, 577]. For instance, the income of the global top 10 percent has been estimated to have been 18 times larger than the income of the global bottom 50 percent in 1820 and 30 times larger by 1860 [103].

As the global poor mostly lived in countries whose economic development was hardly affected by the first industrial revolution during the 19th century, the advancements in operations appear to have offered little economic growth for the world's poorest [71]. The inequalities within industrializing countries were affected less dramatically than GDP over the years 1820–1870 and the most recent evidence suggests that they have been stable or increased [103, 382]. Thus, income improvements were quite broad in these countries, even though the lower-income groups likely benefited less.

7.5.2 Social Inclusion

As is clear from the preceding data on economic growth and inequality, the realized economic growth was not inclusive. More concretely, we have also seen in this chapter how the Indian weavers lost income as volumes and prices decreased. After protest, despair, and suffering, many Indians working in the cotton supply chain, e.g., as spinners or weavers, lost their jobs and welfare [77, 484].

British spinners and weavers protested against the industrialization too. Spinners broke into James Hargreaves house and destroyed his first spinning jenny to protect their work and income [547]. Such destruction would reoccur when Hargreaves set up a nearby spinning workshop. Richard Arkwright also worried about such protests and destruction and set up his first spinning mills at a distance from the regions where home spinning was most actively practiced [625]. While employees appeared willing to defend the factory, he feared protesters from elsewhere enough to additionally purchase a canon and small arms [625].

The spinning mills ended self-employment spinning from home and replaced it with employment in tightly managed cotton mills with 13-hour working days, 6 days a week, mostly populated by women and children. In fact, some machines were designed to be operated by children, as employed by Arkwright from the age of 10 years. The poem opening the chapter provides an impression of how operators experienced their half-day working shifts. The productivity pressures resulted in demanding working conditions for employees and the loss of work and income for self-employed home spinners while enabling Richard Arkwright to collect his famous fortune.

Still, the factory workers were relatively well off in comparison to the workers in the mines. Working days of more than 12 hours became the rule in the mines too, for men, women, and children alike, some of whom were less than 8 years old. The working conditions in the mines were much harsher and brought the risk of death and injury as well as other health risks. The spaces were narrow, sometimes only 75 centimeters high, and the lighting was poor and dangerous, as illustrated in Figure 7.3. Temperatures were high, and conditions were often humid and slippery because of the water entering the mine. It was not uncommon for the air to contain toxic gases and dust and be lacking in oxygen. A working place of deathly occupational hazards, injuries, and diseases [383].

As the demand for coal grew, mining companies extended their operations, setting up larger, more complex mines and extracting from deeper underground where access was more difficult. The workforce increased, organizations formalized control of operations, and the workforce tightened for the purpose of productivity [383]. The workers in the dark underground world of British mining operations,

Figure 7.3. Mining Operations in Great Britain (source Science Photo Library).

with the pressures, norms, and behaviors that developed in this operating context, experienced a life expectancy of 36 years in coal mines and of 29 years in metal mines. Such is against a life expectancy of 62 years for agricultural workers [383].

In 1842, a royal commission report on the employment of women and children in mines caused widespread public dismay at the depths of human degradation that were revealed [433]. Owners were considered to have shown critical lacks of concern and responsibility for the welfare of their workers, and the British parliament forbid women, girls, and boys of age 10 years or younger to work underground [433]. Despite these measures and the enforcement of a number of additional regulations, the yearly mortality remained above 1,000 until 1870, after which it started to decline slowly [433]. Over the same period, British employment in the mines increased from 150,000 to 377,000 [113].

These sad facts about mining operations were not exclusive to British mines during the industrial revolution. Similar harsh working conditions developed in coal and iron mines in other countries in which the industrial revolution advanced, such as the United States, Germany, Belgium, et cetera. Spanish mining in Latin America importantly relied on the use of slavery and coerced labor delivered by the indigenous population [44, 82]. These mining operations and the living conditions of the indigenous workers were lethal to the extent that some mining operations were halted when the locally available workforce had been depleted [82]. The extraordinary and difficult working conditions for miners in Roman times have already been covered in Chapter 5.

Globally, the miners who produced the coal and metals that formed the energy and prime materials feeding the industrial revolution were among those most negatively impacted by the "*war waged on the working population*" at the time [383].

Among the other workers in the operations of the global supply chains and operations of the new industrial era that were excluded from the progress it brought were the sailors working on the ships and the slaves working in the cotton fields. The 12 million slaves brought from Africa to the USA to work in the cotton fields or elsewhere on the new continent are considered to have been worst off—if they survived the trip across the Atlantic Ocean at all [32]. Such is the case despite operating at the source of the cotton supply chain and delivering an essential contribution to the cotton industry in Great Britain and the USA, which formed such an important driver of the industrial revolution [32].

The men operating the ships that sailed the world to connect the newly arisen global supply chains also experienced considerable hardship. The journeys from Europe to the Americas and to India could take many months, during which the crew would live on food brought along, such as salted meat. As the diet lacked vitamin C, the disease scurvy was highly prevalent. In 1497, Vasco de Gama lost 116 of his 170 crew members to scurvy on his historic journey from Europe to

Figure 7.4. Cotton workers in the field, Oklahoma, USA, late 19th century (source National Research Archive, USA).

India and back. This happened despite awareness of the curative effects of citrus fruits [43].

The same difficulties were experienced by the Spanish, Dutch, and British over the next three centuries, during which the preventive effect of citrus fruit consumption became more widely known but not commonly adopted [43]. In 1741, almost 250 years after Vasco de Gama, Captain Anson left Portsmouth with eight ships and 2,000 men. Only 200 of these men were still alive when returning to Portsmouth in 1744, as most others died from scurvy [555]. Scurvy is estimated to have caused the deaths of more than 2 million sailors until 1800 [368].

In 1747, James Lind conducted the first controlled trial in the history of medical sciences when confronted with twelve cases of scurvy as a surgeon on a naval ship. In addition to (or instead of) a common diet, two out of twelve were given vinegar, two other seawaters, et cetera, and two were given two oranges and a lemon each day [228, 555]. As reported by Lind 6 years later, in 1753, the latter two were much improved after 6 days, whereas the others were not [228, 555]. These results were difficult to understand at the time when vitamins were still unknown and have been misinterpreted.

On the positive side, Lind's findings contributed to increased (citrus) fruit intake in the British Royal Navy. Captain Cook did not lose a single crew member on his famous voyage around the world in 1778. The Royal Navy recommended daily lemon consumption for all sailors in 1795 [43, 228]. For commercial ships, such as those operated by traders in cotton and slaves, scurvy prevention measures were not enforced until 1867 by British government regulation (after merchant ship owners had evaded a previous act from 1844) [43, 348]. Many years and men passed between the forced adoption of life-saving scientific evidence and earlier practical understanding of the shipping operations of the supply chains feeding the industrial revolution.

Altogether, the above makes it clear that the industrial revolution was far from socially inclusive. It relied on extreme forms of social exclusion. The defining operations of the cotton fields, the ships, the mines, and the factories relied on the exploitation of men, women, and children for manual labor. This exploitation included slavery and other forms of coerced labor. It caused death, injury, illness, and poverty among workers, while the entrepreneurial engineers and capital providers owning the commercial organizations running these operations accumulated considerable and sometimes extraordinary fame and wealth.

7.5.3 Environmental Protection

Let us start reflecting on the environmental impact of the industrial revolution by reconsidering the Spanish deforestation example addressed when discussing the environmental impact of the agricultural revolution in Chapter 5. The Spanish Armada grew rapidly after the discovery of the Americas, expanding global supply chains for metals, sugar, spices, et cetera. It has been estimated that from the 15th to the 18th centuries, well over 13 million trees have been cut to acquire wood and tar for ship building and to fuel furnaces to produce cannons. This resulted in more than 50,000 hectares of land to be deforested [569]. As accessible oak and pinewood became scarcer in Spain, the Armada turned elsewhere for timber and developed shipyards in its colonies, in particular Cuba [376]. In combination with large-scale sugar plantations and production, the ship building industry had caused the deforestation of major parts of Cuba by the end of the 19th century [236].

Likewise, the Dutch and British colonial operations, and particularly ship building, caused large-scale deforestation. It has been suggested that more than 2 million hectares of teak forest disappeared in Indonesia after the arrival of the Dutch East India Company [68]. The British caused major deforestation of teak forests in India after depleting their own oak forests [212]. The growing British demand for cotton during the 19th century was accompanied by policies to stimulate cotton growing in India, which caused additional deforestation [212]. In the United States, deforestation for the purpose of cotton production changed the ecosystem of the South-East "Cotton Belt" states from their original mixed forest vegetation [66]. As elsewhere, the deforestation caused changes in biodiversity. Moreover, it also destroyed long-existing human niches when the indigenous populations that had lived in these ecosystems for thousands of years were forced to move West.

The combined overall effect of these increases in land use for cash crops such as cotton and wood harvesting—two forms of human ecosystem engineering—increased the net carbon production of these ecosystems and anthropogenic GHG emissions [496]. The use of wood as an energy resource to produce steam and power industrial production operations also negatively impacted the sustainability of human operations during the first industrial revolution.

As the industrial revolution progressed, wood was soon replaced by coal as a main source of energy to produce steam and for metal production (see Figure 7.2). With similar developments taking place in other countries, be it to a lesser extent or at a later stage, the coal-fueled first industrial revolution caused an exponential growth in anthropogenic GHG emissions in general and in CO_2 emissions in particular [514]. Moreover, the negative impacts of coal production, preparation, utilization, and combustion on the environment go far beyond GHG emissions and the resulting global warming. Together, these operations negatively impacted agricultural productivity, water quality, and vegetation; they destroyed existing ecosystems and caused erosion, ground movement, surface deformation, and acid rain [59].

The air pollution caused by coal combustion has negatively impacted many species, among them humankind itself. As we shall see in later chapters, subsequent industrial revolutions and the continued use of coal as an energy source have caused air pollution to be among the world's leading health risks and indeed have caused millions of premature deaths annually in recent centuries. The negative health impacts of coal combustion in general and on the lungs in particular were widely felt and known in 19th-century Great Britain [255]. Such loss of health and resulting loss of labor productivity, combined with other factors such as reduced migration into the most polluted cities, has negatively and significantly fed back into economic growth [255].

As already extensively addressed, the industrial revolution has also importantly depended on iron and iron mining. Iron mining operations may also entail a variety of negative environmental impacts, among which have been and are damage to land, air pollution, water pollution (with hazardous metals such as lead, cyanide, and acid water possibly with toxic metal content), and toxic waste in the form of dust and (illegal) dumping [185]. In addition, iron smelting is very energy-intensive and typically involves coal. The smelting operations release GHGs and may cause acid rains [185].

7.6 Operations Management Perspectives

The industrial revolution advanced on the agricultural revolution described in Chapter 5. Operations were powered by human and animal energy and by the renewable resources of water and wind power when the likes of Vasco de Gama and Columbus sailed out to find new sea routes to India to obtain hand-made agricultural products such as spices and cotton. In present-day supply chain management terminology, the industrial revolution started with a modal shift from land to seaborne operations, for which the wind provided the necessary renewable energy.

The new routes were developed to improve access, reduce cost, and gain better control over the supply chains. The control of overseas operations would soon extend to ownership of the land in the ecosystems in which the corresponding agricultural operations took place. These practices were also adopted for other operations for which the newly entered ecosystems turned out to be resourceful, such as the mines in Latin America. The Spanish and Portuguese managed these operations through the governmental structures of their kingdoms. The British and Dutch depended importantly on private enterprises to run these operations.

The products from overseas were in great demand, and their supply therefore held great potential for profitable operations, as long as production costs were sufficiently low. In addition to taking control over natural resources such as land and mines, the operations management at the source of these new supply chains thus also generated a demand for affordable human labor. The continuous scale-up of operations in response to increasing demand caused to recruit more and more low-cost workers, often in non-sustainable ways, among which were the use of violence, coerced labor, and slavery. The difficulty to attract sailors for the scurvy-plagued naval activities was such in 18th-century Great Britain that abduction was one of the recruitment instruments [460].

All these developments were importantly driven by the human labor intensity of the operations and to address the resulting challenges of controlling costs while scaling up production. The search for efficiency and economies of scale sparked the industrial revolution, which substituted human labor with machines, technology, and capital goods. This replaced human and animal energy as a power source. Initially, by wind and water, as was the case for Arkwright's first cotton mill and the ships supplying the raw cotton, the next threshold step involved the introduction of steam-powered machines for which steam was generated by burning coal. The second and iconic Cromford cotton mill, designed and operated by Arkwright, delivered these improvements for cotton cloth production.

Similar production facilities were simultaneously developed for other textiles and in other industries, such as transportation, where steam-powered locomotives caused revolutionary changes. Operations management thus increasingly regarded the management of the steam-powered machines, which conducted the primary operations of processing inputs into outputs. The workforce was subsequently managed as a function of machine operations. Below, we elaborate on two important corresponding operations management developments from this era of the industrial revolution more closely.

7.6.1 Operations Management for Mass Production

While we have learned about the mass production of pottery in early civilizations in Chapter 5, the mechanization of the industrial revolution increased the volumes obtained in mass production by several orders of magnitude. The mass-produced pottery of the early civilization relied on piece-by-piece manual operations, or "manu-facturing," by craftsmen and women. The mass production operations of the first industrial revolution were powered by steam and used machines that allowed lesser-skilled workers, such as children who received little training, to operate machines that were hundreds of times more effective than skilled human operators. One may argue that the word "manu-facturing" does not really apply to these operations, even though the industrial revolution is commonly associated with the invention of high-volume manufacturing. Plant designs for low variety, high-volume production, such as the design of the Cromfort cotton mill, formed the first **continuous process** layouts populating the lower right-hand corner of the product process matrix in Figure 5.3 [510].

From an operations management perspective, the high-volume manufacturing process and layout designs were complicated by the spacious demands of the water- and steam-based power systems. These systems used networks of wheels and belts that emanated from a central source (such as a water wheel), and the machines had to be located in alignment with this energy delivery system. To avoid long (horizontal) distances for the energy transportation system, the factories typically occupied several stories, which implied inefficient vertical transportation of raw materials and work in progress.

The energy system was either on or off. When switched on, it continuously kept the system of wheels and belts in motion and was hardly allowed to switch individual machines on and off. This necessitated the machines to become the basis of operations management. The workers began to operate the continuously running steam-powered machines. This explains, for instance, the 12-hour shifts and an emphasis on punctual, compliant with operating standards, and responsive to the needs of the machines. It required discipline and humility to work in the function of the system, however demanding or monotonous the operations, as vividly expressed in the poem opening the chapter. The poem was written during the Lancashire cotton famine, during which American cotton was hardly available because of the civil war and British cotton mills had turned to processing Surat cotton which was more difficult to process and more likely to break. An individual operating mishap by any of the workers could affect the mass production operations of the plant as a whole and significantly reduce production output, revenues, and costs. Correspondingly, controlling a lowly skilled workforce to be disciplined, respecting the organizational hierarchy, and executing assigned tasks timely and according to operating standards became a key operations management focus of high-volume, low-cost manufacturing.

The substitution of skilled, manual, human labor by machines generated cost and quality advances, growth in demand, and corresponding growth in production volumes. Somewhat ironically, perhaps, this ultimately caused the number of low-skilled workers required to operate labor-extensive steam-powered operations to outnumber, by far, the initial population of skilled craftsmen and women working in the human labor-intensive operations that were made obsolete by the industrial revolution.

7.6.2 Standardization through Interchangeable Parts

In 18th-century Europe, the adoption of newly designed machines in manufacturing advanced not only in the continuous processing of agricultural products but also in discrete manufacturing operations of parts (such as gears) for tools and appliances such as clocks, watches, and guns [618]. The novel idea of having machines produce parts in high volumes according to standard specifications to later assemble products from arbitrarily selected uniform parts naturally supported a switch toward mechanized manufacturing. It avoided the natural variation of human production, however skilled the craftsmen and women. Moreover, it made recombination of parts and product repair easier.

The idea of "interchangeable parts" or the "*uniformity principle*" became important in the purchasing of muskets by the government of the United States in the late 18th century. The advancements made for the production of muskets using interchangeable parts by the US national armories led Eli Whitney and other contractors to the design of new machines, tools, and processes [618]. These government contracts regarded several thousands of muskets of already existing models. Nevertheless, the suppliers experienced difficulties producing the requested volumes while satisfying the interchangeability requirements [276, 618]. It would take many decades of subsequent costly advancements in methods and machines to achieve interchangeability, even when allowing for some refitting [276, 618].

In 1860, more than half a century later, the volumes of guns manufactured in the USA for private use were "*several times larger than that produced or procured by the military*" [276]. While this implies additional opportunity for economies of scale to earn back investments in methods and machines for standardization in the form of interchangeable parts, such practices were not fully adopted in this private sector in 1860 (and would still not be more than a century later) [276].

Still, the foundations for an "*American system of manufacturing*" were established in the first quarter of the 19th century. An important founding principle of this system was the standardization of parts, which were mass produced by steam-powered machinery designed for specific operational purposes [618]. The next chapter shows how these principles of operations management were further elaborated in the second industrial revolution and, for instance, adopted in the assembly line.

DOI: 10.1561/9781638282853.ch8

Chapter 8

Engines and Electricity

Opportunity is missed by most people because it is dressed in overalls and looks like work.

Thomas Edison

8.1 The Operations of Invention

The British scientist Isaac Newton laid the foundations for many of the mechanical laws adopted by the engineers of the industrial revolution. Newton's scientific operations required a combination of skills, among them creativity. Myth has it that his creativity was stimulated by an apple falling on his head from the tree in his garden while he was taking a nap [189]. This incident supposedly helped him to conjure up the law of gravity.

Newton was not alone in his use of the operation "napping" to be more creative and productive. Other well-known operational nappers are Leonardo da Vinci, Salvador Dali, and Thomas Edison [592].

Thomas Edison personifies the many and important 19th-century industrial and societal innovations of the United States of America that define the second industrial revolution. He allegedly developed standard operating procedures to maximize the effectiveness of napping as an invention. The standard operating procedure was to sit down for the nap in an armchair with ball bearings in his right hand and a metal pan on the floor right next to the armchair. Whenever he would fall asleep and start to dream, his arm muscles relaxed, causing the ball bearings to fall and hit the metal pan on the ground. The noise of the ball bearings hitting the metal pan then ended Edison's nap. The purpose of this carefully designed layout and process was to vividly recall the interrupted dream, which might present creative solutions for unresolved problems or other innovations [592]. Edison's operations of napping and waking up helped the second industrial revolution come about.

In the second half of the 19th century, Edison was one of many American men and women of various socioeconomic backgrounds working hard to capture their dreams and drive the second industrial revolution [297]. As it will be impossible to give a comprehensive, or even representative, overview of the many efforts these men and women conducted, this chapter highlights some exemplary developments of the second industrial revolution, among which are the contributions from Edison's Menlo Park. In addition, it hopefully pays brief, yet due, credit to other essential innovators of the second industrial revolution from the US and elsewhere.

8.2 Edison's Invention Factory

Edison's operations of invention were organized far beyond the design of the single, personal process for napping and waking up. Edison built and led the "*invention factory*" at Menlo Park. The research and development facility at Menlo Park has produced important innovations in telegraphy, audio, film, electric lighting, electric power systems, and more [283]. The factory was a blend of research laboratories and small workshops in which teams of engineers would collaborate with a variety of expert practitioners [283].

Electric light came into existence through the contributions of a variety of inventors in several countries in the early 19th century [515]. The first patents for an incandescent lamp, or light bulb, were granted in 1841 in the United Kingdom. The first patents for an arc light were granted in 1845 in the United States of America. The first commercial uses date back to 1862, and arc lights were already illuminating the boulevards of Paris and of Broadway in New York in 1879 when Thomas Edison came to the scene with the first light bulb developed at Menlo Park [515].

Edison also worked on the electricity supply further upstream in the electric light value chain. He developed electricity supply networks, which operated on the basis of a direct current (DC) system. Edison filed patents for these inventions around 1879. Westinghouse and the Thomson-Houston Electric Company were Edison's main competitors in operating energy supply networks. Thus, there were alternatives for both the light bulb and the electricity supply, and competition would develop over the next few years involving patents, lawyers, investors, mergers, politics, marketing, and public relations, in addition to further product improvements realized by Edison's extensive team working in the invention factory in Menlo Park.

In contrast to Edison's General Electric Company, the networks of Westinghouse and the Thomson-Houston Electric Company used alternating current (AC). AC is much better and more economically enabled for longer-distance electricity networks. In 1892, Edison's General Electric merged with the Thomson-Houston Electric Company to form General Electric. Upon its creation, General Electric was by far the largest electricity provider and would remain so for quite some time, powering lighting in streets, homes, and businesses.

8.3 Electricity and the Electric Engine

Access to electricity as a new source of power operations brought many changes to the way we live that continue until today. The convenience of electric lighting at home was further enhanced by the introduction of wall light switches (substituting the central chord that was hard to find in the dark) by engineer and psychologist Lilian Gilbreth [238]. A later innovation in household operations contributed by Lilian Gilbreth is the circular workplace, later translated to the *kitchen triangle*, a layout designed for efficiency in kitchen operations. The triangle included the sink, the stove, and the electrically powered refrigerator. The refrigerator is one of the products of the second industrial revolution, with a sustained impact on how we live. For the refrigerator door, Gilbreth invented the interior shelves [238]. Together with her husband, Frank Gilbreth, Lilian worked as one of the first operations management consultants in industry and transferred industrial methods and principles to the household operations of life. Their contributions are revisited in the final section of this chapter.

The networks of General Electric and others also provided electricity to business customers. The business interest in electricity was not so much rooted in access to electric lighting but mostly in the utilization of another technological innovation that defines the second industrial revolution, the electric engine. The electric engine developed over several decades with contributions from a variety of inventors, sometimes building on each other's work, sometimes working in parallel, or even ignorant of progress elsewhere [572]. Altogether, electric motors started to be used in a variety of operations before the end of the 19th century [572]. More and more manufacturing companies contracted electricity, and network capacity grew rapidly to match the increase in demand [165].

The initial applications of electric engines often involved the replacement of water power and (coal-based) steam power. An early large-scale application of such electrification, for instance, occurred in the cotton supply chain when the Columbia

Cotton Mill in Colombia, South Carolina, replaced its steam (coal) power system with 17 electric engines. Likewise, a newly constructed 10,000-spindle cotton plant in Athens, Georgia, was powered electrically in 1895 [165].

Electric engines were preferable to steam engines in manufacturing operations for a number of reasons. First, electricity could be purchased from providers, thus eliminating the need to produce energy through water, coal, or otherwise. This also avoided the need to invest capital in energy production. Second, the electric engines were considerably more efficient and reduced the required energy supply [165]. The efficiency gains would be further enlarged by the possibility to switch machines off and on as needed, where coal-steam systems would often operate at constant capacity, regardless of use [165]. Together, these gains could lead to capital and energy cost reductions of 70 percent or more.

Third, as explained in Chapter 7, the water and steam-powered plants were operated using a system of shafts and belts that had many implications for plant layout. Hence, today's plant layout would be partially organized around power supply instead of best facilitating the workflow of the primary process. As technology advanced and more powerful electric machines were introduced, longer single-story plant designs appeared, which adopted a product layout following the sequence of production operations for a single product or set of closely related products [254]. Such investments became more attractive as access to electric power provided through the utility network improved. Especially so for low-variety, high-volume production that yielded sufficient economies of scale.

Product layouts were soon adopted in various industries. For example, let us briefly consider the sugar industry. Early cane sugar production in the United States and elsewhere would often be done in a small-scale workshop at the plantation after the harvest was completed. The process involved a sequence of kettles grouped together, and the operation required skillful operators to produce sugar crystals from the canes [259]. By 1830, a couple of decades after introducing cane sugar farming to Louisiana, these plantation-based workshops together produced 33,000 tons of sugar annually [259]. Less than a century later, and after the introduction of beet sugar production, there were more than a hundred large-scale plants in the United States with a product layout that (on average) were capacitated to produce this quantity of 33,000 tons per month [30].

The growth in volume of sugar production by three orders of magnitude shows the potential of efficient industrial mass production to satisfy high demands. The reductions in investment costs and operating costs brought by electric machines would also make the use of machines also attractive and accessible for smaller manufacturers, replacing manual or animal labor. By the turn of the century, half of the American manufacturers were using electricity. By 1930, more than 90 percent of the American manufacturers used electricity to power operations [165].

8.4 The Car Manufacturing Assembly Line

Like Thomas Edison, Henry Ford was the son of a farmer and personally embodied the transition from agriculture to industry. According to the Henry Ford Organisation: "*Henry Ford did not invent the automobile. He didn't even invent the assembly line. But more than any other single individual, he was responsible for transforming the automobile from an invention of unknown utility into an innovation that profoundly shaped the 20th century and continues to affect our lives today*" [203].

In 1904, the Ford Motor Company set up the first plant to produce cars in Detroit. This Piquette Avenue plant was a multi-story building that largely adopted the aforementioned central wheel mill layout [438]. The production volumes of the various models produced at the Piquette Avenue plant were modest. For instance, the Ford Motor Company produced 500 cars of its second model, model B, over the period 1904–1906 in the plant [14].

In pursuit of larger volumes and lower costs, the company worked on the design of the new Highland Park plant. This single-story plant included an electric power generation unit and used electricity to power machines and provide lighting in areas further away from the windows [15, 408]. The design adopted a product layout in which machines were located to best facilitate the sequence of automotive manufacturing operations.

Ford had learned about product layouts from the meat processing industry, in which operators worked along a single work flow path below an overhead rail conveyor belt transporting pig carcasses. Each of the operators repetitively performed a relatively low-skilled, specialized (cutting) task according to their position alongside the belt [23]. In 1859, Cincinnati was well known for its pork meat industry, and the workers reportedly "*chopped a hog*" every 35 seconds [23]. Overhead rail conveyors were common in the city's slaughterhouses by then [23]. Henry Ford allegedly observed this highly standardized and structured disassembly process several decades later in Chicago, which was the largest meat packing hub in the early 20th century. The design of this efficient and high-volume production process inspired the product layout-based design for the Highland Park plant, where the Ford Motor Company planned to produce Model T Fords at low costs and in high volumes.

The Ford Motor Company opened the new Highland Park Plant on January 1, 1910 [408]. Next to the factory itself and the power generation unit, Highland Park hosted a foundry and component factories [15]. Along the linearly organized manufacturing process, standardized components (interchangeable parts) were assembled using standardized methods that could be learned quickly by lowly skilled operators, even with limited mastery of the English language. By 1913, this linear assembly layout had gradually evolved to incorporate a constant-pace moving

Figure 8.1. The Highland Park assembly line, source Ford Media Center.

assembly line to improve efficiency [408]. Figure 8.1 provides an impression of the assembly line at Highland Park and the component assembly lines feeding it. Moving assembly lines are still the norm in automotive manufacturing today and commonplace in many other mass production industries as well.

The moving assembly line forms a threshold innovation in the operations of the second industrial revolution. To appreciate how it impacted the way humankind lives and works, let us quantify some of the developments driven by moving assembly line-based car manufacturing. In 1900, 1 in every 10,000 inhabitants of the United States owned a car, for a total of around 8,000 cars [5, 100]. By 1909, the year before the Highland Park plant started the Model T production, the number of cars had risen to 3.5 per thousand [5, 100].

While the original operating model relied almost entirely on the assembly of externally produced components, the company increasingly produced components in-house and at a lower cost. Moreover, it managed to do so while doubling labor productivity from 1909 to 1916 [606]. As a result, in-house labor costs per car remained fairly stable between 60 and 70 USD, despite an increase in daily wages from 2.37 USD to 5 USD in 1914 [466, 606]. The relentless improvements to the manufacturing operations caused the production cost of a Model T car to decrease by more than 50 percent, from 560 to 265 USD [606].

Driven by all these operational improvements, the yearly production volumes of the Model T Ford would increase from 14,000 in 1909 to 585,000 in 1919 and peak around 2 million in 1923. In 1923, the 2 million T Fords produced formed more than half of total car production in the USA [5, 275]. By the time the Ford Motor Company ceased production of the Model T in 1927, it had produced 14.7 million of them in the 18-year period since commencing. This number is even more remarkable in view of the total number of around 20 million registered cars in the United States in 1927 [5]. Car ownership had increased from 1 in 10,000 in 1900 to 1 in 5 in 1927 [5, 100]. A revolution in the way Americans lived had been accomplished within 30 years.

The electrically powered moving assembly line enabled this societal transformation by providing affordable access to cars. Cars came into existence because of a complementary and related invention: the internal combustion engine. Like electric engines, internal combustion engines are an advancement over (external) steam engines, and a variety of innovations contributed to their advancements over the course of the 19th century. Important final contributions for the development of liquid fuel-consuming internal combustion engines to propel automobiles were made in Germany by innovators with names such as Gottlieb Daimler, Karl Benz, and Rudolph Diesel [310]. The cars greatly increased the radius of action of their owners and their families and changed mobility behaviors in support of more dispersed patterns of settlement, socialization, and work [158].

8.5 Oil and Gas

Hunter-gatherers burned animal fat in stone lamps before the agricultural revolution. Early agricultural societies witnessed a widespread uptake of pottery lamps fueled by organic oils such as olive oil and sesame oil. The use of petroleum-related resources such as bitumen for lighting likely dates back more than 3,000 years [257, 339]. This was not the first use of petroleum, however. The petroleum-related product, natural asphalt, was already used for the construction of houses and ships in Mesopotamia at the onset of the agricultural revolution, 11,000 years before present [122]. Asphalt was first transported from the Red Sea to Egypt no later than 6,000 years ago as a resource for a variety of operations, among which, eventually, embalming [122].

The first reports of the use of gas and oil in China date from around 3,000 years before present. Moreover, it is from China that the earliest drilling for oil and gas at greater depths (more than 100 meters) is reported, well over 2,000 years before present [195].

More recently, oil wells were drilled in Baku, Azerbaijan, to win oil as a fuel for lighting in 1847. The presence of oil in Pennsylvania, United States of America, was well known by that time, and small-scale use for purposes other than energy occurred. "*Petroleum was still in some ways a resource in need of an application*" on the American continent halfway through the 19th century [60]. This soon changed when oil was considered an alternative for kerosene produced from charcoal to fuel lamps. The operations of producing light from oil were less costly than the common practice of producing it from charcoal [60]. This development sparked a revolutionary development in oil winning, with a lasting impact on the way we live and work, as well as on planet Earth.

Over the next decades, applications of oil beyond serving as an energy resource for lighting developed. Well-known important applications are the use of oil as fuel for combustion engines in transportation, especially in cars, direct use to power industrial machines, and combustion engines that produce electricity. The growth of automobile volumes from thousands to millions in a few decades was one of the developments driving the growth in demand for oil. The demand for oil was further stimulated by the relative advantage of oil and gas in energy production operations, causing oil to increasingly replace coal (and wood) for heating and for electricity production [525]. Altogether, the demand for oil rose dramatically in a matter of decades, as illustrated in Figure 8.3.

The production of oil would dramatically change work and life in Pennsylvania. Boomtowns and cities in which oil was omnipresent sprang up in the 1860s. Towns with names such as Pithole, Oil City, and Titus grew from virtually zero inhabitants to more than 10,000 inhabitants within a year [60, 529]. In addition, the Pennsylvanian oil boom gave rise to forests of derricks occupying many miles of the valleys of Oil Creek, Cherry Run, and Cherry Tree Run, as illustrated in Figure 8.2 [60]. Franc B. Wilkie, who traveled through the area in 1865, wrote the

Figure 8.2. Early oil field in Oil Creek, Pennsylvania, around 1860, data source: Pennsylvania Historical Collection and Museum Commission, Drake Well Museum Collection.

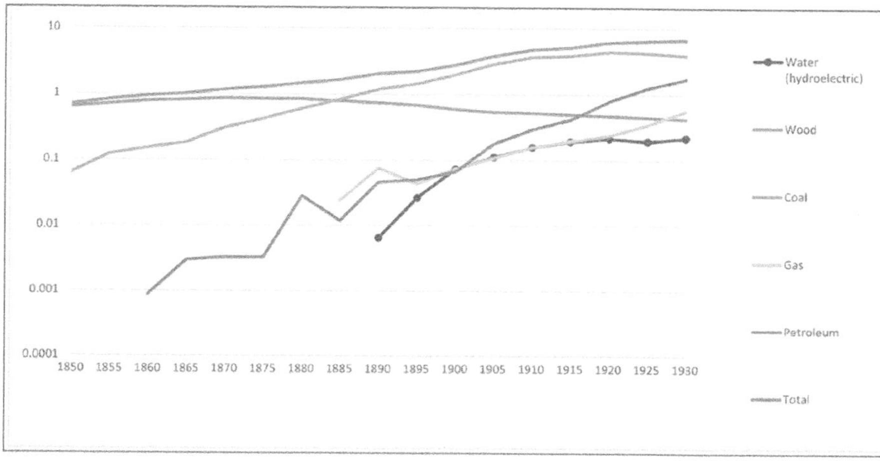

Figure 8.3. Energy use by source in the USA, 1850–1930 in PetaWatt Hours (logarithmic scale), data source [174].

following about Oil City: "*one sees little but oil save mud, and even this has none of the stickiness of usual mud, owing to the universal presence of oil. Wagons in endless length pass southward loaded with oil; the air is tainted with oil. The refineries are blue with it...*" [60].

The Pennsylvanian oil boom formed the start of a new industry whose operations consisted of the winning of oil and gas. While the Pennsylvanian oil cities would be abandoned within years, after the oil reserves appeared depleted, the industry had already spread from its Pennsylvanian origins across the United States of America. As demand for oil grew and transportation networks by rail and pipelines developed, crude oil started to be transported to refineries located closer to the end-user markets. New York City, which already had plants to produce kerosine from coal and hosted more than 5,000 factories by 1860, developed into the city with the largest oil refineries by the mid-1880s [279]. Initially, some of these refineries developed in Manhattan, Queens, and Brooklyn. Soon, however, Northern New Jersey became the center of refinery operations, together with Newtown Creek, where "*John D. Rockefeller decided to concentrate the operations of the Standard Oil Company*" [279].

As is well known, oil and gas-winning operations would spread further across planet Earth in the 20th century. Technological innovations in oil drilling further stimulated these developments as they enabled larger-scale winning of oil and gas at lower costs. In the first half of the 20th century, energy use and the share of oil and gas would grow particularly rapidly in the USA. In 1860, when oil was first recorded as an energy resource, wood was still the main energy resource in the USA, while coal was rapidly gaining relative importance. Coal would surpass wood to become the dominant source by 1885, as illustrated in Figure 8.3. The same figure, which has a logarithmic scale on the *y*-axis, shows that oil and gas made quite modest contributions to total energy production at that time.

The relative contribution of coal to the total energy of coal peaked at 76.8 percent in 1910, the year in which the Highland Park plant started operations. By 1910, energy supplied by coal stabilized around 4 petawatt hours per year, while petroleum and gas continued to grow exponentially [174]. By 1930, coal had formed 64 percent out of a total of 6.23 petawatt hours, of which oil formed 27 percent, while the remaining 9 percent was largely from gas [174]. Over the same period, energy from alternative sources such as wood and hydroelectric power remained stable, just above 0.6 petawatt hours [174]. Altogether, yearly energy production (and consumption) in the USA thus grew ten-fold over the period 1850–1930, from 0.69 petawatt hours to 6.94 petawatt hours.

Global energy consumption from fossil fuels amounted to 6.1 petawatt hours in 1900, of which 95 percent was from coal and 5 percent from oil and gas [525]. Thus, the USA consumed more than one-third of global energy from fossil fuels at the turn of the century, while hosting less than 5 percent of the global population. These numbers evidence the lead of the United States in the second industrial revolution.

By 1930, global energy consumption from fossil fuels had more than doubled to 13.1 petawatt hours, roughly half of which was consumed in the United States [174, 525]. The share of coal in the global production was 80 percent, and oil and gas together contributed 19 percent of global energy. Globally, electricity provided a modest 1.5 percent of total energy consumption, at 180 terawatt hours. Throughout the first half of the 20th century, the United States would generate and consume more than half of the global electricity for its operations of work and life [101, 525].

The rise of the manufacturing industry in developed countries significantly impacted the ways of working of their populations. By the end of the 19th century, almost 59.4 percent of the workforce of developed countries worked in agriculture and 16.8 percent in manufacturing [34, 242]. While the majority of the workforce in many developed countries had been engaged in agricultural operations for thousands of years since the spread of the agricultural revolution, this would no longer be the case from 1930 onward [34, 242]. The growth of manufacturing productivity in developed countries over this time period was mostly driven by deploying more technology. The percentage of the workforce employed in manufacturing grew slowly yet steadily from 16.8 to 18.6 percent by 1930 [34].

Throughout the same period, more than 75 percent of the workforce in developing countries worked in agriculture, while manufacturing operations formed the place of work for less than 10 percent of the working population in these countries [34]. Chapter 9 covers how the service sector completes the employment picture.

8.6 Sustainability of the 2nd Industrial Revolution

8.6.1 Economic Development

The per capita GDP of the United States doubled between 1830 and 1880 and doubled again until 1930, when the great depression set in [65, 97]. As the population almost grew by a factor of 10 over the same period, GDP increased by a factor of almost 40 within a century of the continued industrial revolution. Over the same period, the per capita GDP in industrializing Western Europe tripled, while population growth was below 100 percent [65].

The global population doubled over the 1820–1930 period, from slightly over 1 billion to slightly over 2 billion [357]. The growth of global real GDP per capita is depicted in Figure 8.4. This global real GDP per capita growth and population growth together yielded a five-fold real GDP growth for planet Earth between 1820 and 1930. As is clear from the above, the developed regions, especially the USA, industrialized their operations at a higher rate and experienced above-average real per capita GDP growth over this period.

The increased access to fossil fuels and their deployment as energy resources for the innovative operations of the first two industrial revolutions likely have crucially determined the economic growth of this era [234, 544]. The second industrial revolution made coal-fueled steam more easily accessible and applicable as an energy resource by converting it to electricity. Moreover, the internal combustion engine provided energy from oil and gas to power the operations of humankind at work, at home, and in between, thus opening up a new source of economic growth.

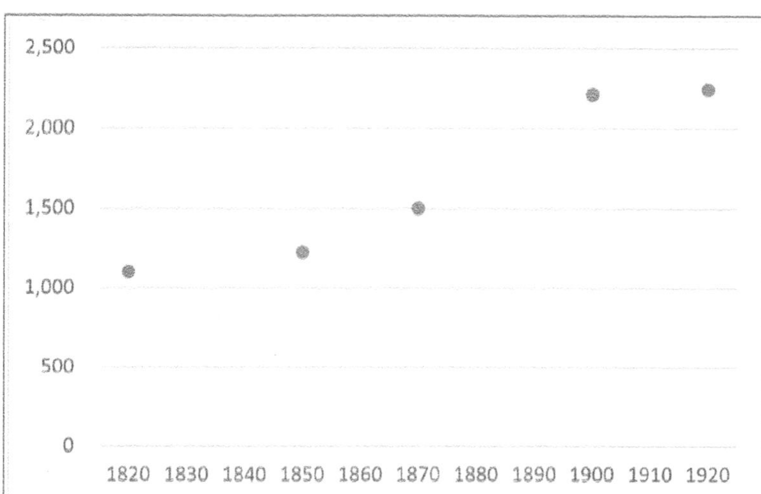

Figure 8.4. Real global GDP per capita, 1820–1930 in 2011 USD, data source [357].

The innovation in operations to manufacture oil-powered automobiles with internal combustion engines on an electricity-powered moving assembly line is thus emblematic of the second industrial revolution.

The automobiles produced on assembly lines also provided humankind with much increased mobility for the daily operations of living. Subsequently, assembly line operations would also provide humankind with refrigerators, hair dryers, telephones, washing machines, dish washers, televisions, et cetera. Eventually, even airplanes were and are made on assembly lines. Each of these machines uses electricity, oil, or gas and free up human energy to conduct other operations that may yield further economic development.

While the overall economic growth that developed countries achieved over the period 1780–1930 has been unprecedented and benefited large subpopulations of an overall increasing population, the economic advancements at local levels and within shorter time frames were not always sustainable. We have already discussed the abandoned Pennsylvanian boomtowns. While the global automotive industry has developed quite steadily, it has brought sharp rises and falls in economic activity and population numbers locally over the course of the 20th century. Detroit, nicknamed Motor City, hosted a population of less than half a million in 1910 and crossed the million mark by 1920 [267]. Its population would eventually exceed 1.8 million by 1950, yet return to below 1 million before the end of the century as much of the automotive industry in Detroit ceased to be competitive and ceased operations [267].

Such developments in manufacturing operations also lead us back to the question of whether economic growth has been equitable and whether the second industrial revolution has brought economic growth for the poor. A classic economic view of the time posits that higher wages for the poor, low-skilled workers in manufacturing diminish the opportunity for the wealthy industry owners to invest in technological advancement. This view would imply there is a trade-off between economic growth and equity [343]. This school of thought has been invalidated, and evidence suggests that inequality may actually hamper economic growth [343, 444]. Indeed, the initial increase in income inequalities generated by the first industrial revolution in the UK was partially reversed in the second half of the 19th century. In the US, these reductions in equality happened at a later stage for a variety of reasons, such as the continued inflow of a low-skilled immigrant labor force [343]. Current understanding suggests that the technological advances realized in developed countries can jointly promote economic growth and equality within countries, as has been the case for the US in the 20th century until the 1970s [344].

On a global level, it appears that the within-country inequalities, which rose steadily (but more slowly than between-country inequalities), peaked around the turn of the century or shortly after (depending on measurement instruments), after which they returned to lower levels [103]. Such is the case while real per capita GDP grew rapidly in countries experiencing the 2nd industrial revolution and typically much more slowly in countries without industrialization. This implies that the lower income groups in industrialized countries benefited more than proportionally from per capita GDP growth in the first decades of the 20th century, while the between-country inequalities determining the inequality levels with the global poor continued to grow [382]. The global number of poor has been estimated to have been quite stable over the period 1820–1920 covering the first two industrial revolutions [71]. Altogether, global inequalities were at least as big or larger than the inequalities within countries and regions over this period [103, 577].

8.6.2 Social Inclusion

Especially in the early stages of the second industrial revolution, the business and operating models that emerged were far from inclusive. They often used unskilled human workers for physically demanding repetitive jobs of 12 hours per day. The jobs were designed in function of the machines and, for instance, followed the rhythm of the moving assembly line. In many manufacturing contexts, unionization was not permitted, and leaving the company was often the only alternative for employees who sought to improve their situation. At Ford's Highland Park plant, "*any worker who refused instructions was sacked on the spot*" [605]. Moreover, jobs could easily disappear because of further technological innovation and process optimization. "*Demanning*" was an ongoing operations management process at the Ford Motor Company, in the never-ending search for efficiency improvement. The employee turnover rate was as high as 70 percent at Ford Motor Company in 1913 (before doubling the salaries) [605].

The harsh conditions in which the blue-collar workers of the industrial revolution operated extended from work to life. Not in the least because of the many and long working days and the poor wages. Moreover, industrial jobs in manufacturing, mining, construction, et cetera, exposed the workforce to occupational hazards such as air pollution, toxic materials, mechanical hazards, et cetera [59, 230]. Such occupational hazards were hardly recognized or compensated for and remain a concern to date [230].

On the positive side, and as illustrated by Ford's doubling of the salaries, the societal and economic developments as importantly driven by the industrial revolutions moved toward more egalitarian income distributions in the USA for many decades of the 20th century [232]. Other industrialized countries, mostly Western European countries, have commonly experienced similar increases in income equality and welfare for all income groups around the same time or afterward.

The inclusiveness of these income advancements was not without limitations. For example, the participation of women in the workforce was limited and would only start to develop more equally with the growth in the number of service jobs discussed in Chapter 9. Racial disparities have persisted throughout the ongoing industrial revolutions (and, in fact, until today) in the USA and many other countries. Despite the abolishment of slavery in the second half of the 19th century, better-paid jobs may be more difficult to obtain depending on race, and race continues to form a determinant of salary [545]. Likewise, migrant status has a long history of being negatively associated with employment in more highly paid jobs and with lower salaries for the same work [545].

Lastly, it should be observed that as the USA industrialized and "developed," there was less and less territory for the native American population, which depended importantly on hunting and gathering operations, sometimes in combination with agricultural operations. Through conflict and forced migrations, many of which took place during the industrial revolutions, the size of the ecosystems they operated in was reduced to 1 percent of their original size [192]. Moreover, this 1 percent of space was often in a different, more arid location [192], which was less amenable to their operations. Together with the population relocation operations themselves, this resulted in high mortality rates (see, for instance, [557]).

In search of valuable energy resources to fuel industrialization, the location of oil and access to oil and other natural resources (such as gold) have repeatedly played an important role in land losses and relocation decisions and have resulted in a strategic exclusion of native Americans to participate in the emerging energy and industrial sectors by the US government [192]. The resulting loss in returns from hunting and gathering operations went by and large unnoticed in national accounts, which consider agricultural, fishing, and forestry as the primary sector but leave (informal) hunting and gathering operations unaccounted for.

8.6.3 Environmental Protection

When discussing the environmental impact of the first industrial revolution, we have already highlighted the negative effects of mining, both for metals and for coal, and coal combustion. As is apparent from Figure 8.3, these effects expanded during the second industrial revolution. Moreover, they were complemented by the environmental impact of oil and gas winning and combustion. Altogether, these fossil fuel-driven operations caused an exponential growth in CO_2 emissions over the years 1850–1930. Over this period, US CO_2 emissions roughly grew a hundredfold, as displayed in Figure 8.5.

In addition to the atmospheric damage, the winning and use of oil and gas also had harmful effects on water and land. While the Pennsylvanian oil wells have

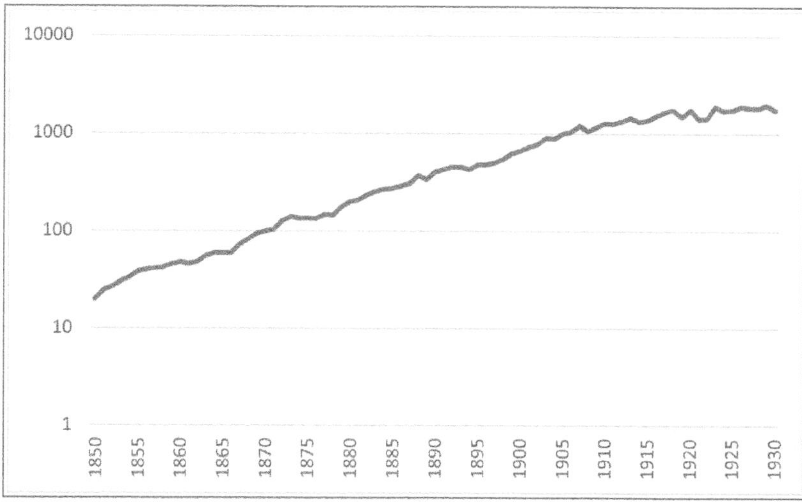

Figure 8.5. US CO_2 emissions in 1850–1930 in MtCO_2, data source [623].

mostly long dried up since, water and land are still polluted to this date. In New York, the pollution caused by the large-scale oil refineries, which were conveniently positioned on the shores opposite Manhattan, reduced the populations of oysters and other shellfish in the local ecosystems [279]. By the turn of the 19th century, biodiversity loss had advanced to the extent that fish had disappeared from the kills around Manhattan. In addition, birds and species living on land suffered from oil waste from plants and visiting boats [279]. It would take until the 1970s before the ecosystem started to be restored in New York.

The briefly summarized impact on land and water ecosystems of the oil industry in the city of New York serves as an example for similar severe, negative impacts in other locations in the United States and elsewhere. The negative impact of the oil industry has remained a global concern to date. This negative impact is further exacerbated by the air pollution caused by the oil and gas industry. In addition to the CO_2 emissions already presented above, these negative environmental impacts, for instance, include other GHGs such as CH_4 and N_2O. In general, the following four categories of air pollutants from winning and use of fossil fuels can be distinguished [292]:

1. Gaseous pollutants, among which are SO_2, N_2O, CO_2, ozone, and volatile organic compounds,
2. Persistent organic pollutants such as the dioxins,
3. Heavy metals such as lead and mercury, and
4. Particulate matter (commonly distinguished are PM10 and PM2.5).

These forms of pollution result from gas leakage and venting during extraction, processing and transportation, flaring, and oil processing [12]. The pollution impacts ground-level ozone, may cause smog, and causes a loss of health for humankind and other species [12]. Among the adverse health effects are increases in hospital admissions and mortality, respiratory illnesses, oncological conditions, cardiovascular illnesses, illnesses of the nervous and urinary systems, and birth defects [292].

8.7 Operations Management Perspectives

The use of coal, oil, and gas to produce electricity as an energy source to subsequently power light bulbs, assembly lines, and a wide variety of other machines formed the core of the second industrial revolution. These machines and devices are new types of resources in the operations of work and life that are not naturally present in the ecosystems in which the humans operating them live. The design, production, and operation of such non-naturally occurring resources that were predominantly powered by nonhuman energy sources were not entirely new. In previous chapters, we have already seen wind mills, sailing ships, water frames, and steam engines (Chapters 5, 7).

The impact of the key advancements in fossil fuel-based energy supply of the 2nd industrial revolution for electricity production, combustion engines, and otherwise is clearly visualized in Figure 8.3. It shows that energy use grew a hundredfold between 1850 and 1930, while renewable energy made a very marginal contribution.

Obviously, this increase in energy use for operations by two orders of magnitude and the corresponding use of machines and other devices powered by these new forms of energy changed the management of operations completely. In fact, many of the presently recognized principles of operations management stem from this era, as covered in more detail below. Operations management developed and matured with a focus on the management of the machines that were powered by coal, oil, and gas, either directly or indirectly in the form of electricity.

8.7.1 The Birth of the Operations Management Discipline

Steam engines were pivotal in the transformation of the calories contained in coal, oil, and gas into machine operations. While coal, oil, and gas were increasingly available, however, the inefficient energy losses of electricity-producing steam plants, and particularly steam engines, formed a concern. Thus, a committee of the American Council of Civil Engineers looked into inefficiencies and reported in 1898

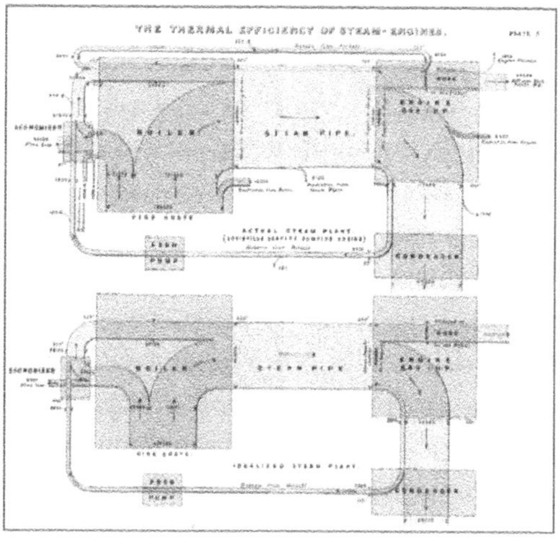

Figure 8.6. The original Sankey diagram, data source [294].

that *"it is also especially useful to ascertain the exact amount of those losses which are inevitable according to the laws of Nature, and to distinguish them from other losses, also to some extent inevitable but yet capable of reduction by improvement in design or in material*" [294]. The report presented the Sankey diagram, a novel model to visualize the energy production operations system and the energy flows therein that was named after the second author (see Figure 8.6 [294]).

The bottom part of Figure 8.6 may best be read from the top left, where we find the boiler. The thick arrow to the right represents the total thermal energy (in British Thermal Units [BTU]) it produces, which flows through the steam pipes toward the engine. In the engine, we see that most of the thermal energy produced flows downward, where it goes to the condensor or feeds back into the boiler. Only 28.5 percent of the thermal energy flowing into the engine flows further to the right, as depicted by the arrow work. This 28.5 percent represents the successfully transformed energy actually powering the engine.

Thus, the Sankey diagram reflects that the majority of the thermal energy produced is not effectively powering the engine for which it was produced. This message is even stronger when realizing that the bottom half of Figure 8.6 refers to an "*idealized steam plant.*" The thermal energy flows for an actual steam plant are depicted in the top half of Figure 8.6 and have a thermal efficiency of less than 19 percent, roughly one third less than the idealized steam plant. Thus, the Sankey diagram identifies opportunities to gain energy efficiency in the design of the idealized engine as well as in closing the gap between the efficiencies of the idealized and actual plants.

Sankey diagrams continue to be used and are instrumental to understanding the present global sustainability challenges, as shown in Chapters 10, 11. We will see that they can be used to model a variety of flows, and their use is not limited to (thermal) energy.

While Sankey and the American Council of Civil Engineers considered the operational efficiency of machines, much of the operations management attention was alternatively focused on the human resources operating between the machines. The design and control of the operations conducted by "*workmen*" or "*workers*" to avoid ineffectiveness and inefficiency of operations became a main thread of the second industrial revolution [202, 551]. This viewpoint was innovative because, as Frederick W. Taylor put it in 1911, "*We can see and feel the waste of material things. Awkward, inefficient, or ill-directed movements of men, however, leave nothing visible or tangible behind them*" [551]. Taylor argued that human labor inefficiencies typically cause human resource productivity to be between one-third and one-half of its potential [551].

Taylor developed "*Scientific Management*" to remedy these human labor inefficiencies [551]. The first principle of his scientific management was for tasks and operations to be scientifically designed so as to maximize the long-term productivity of employees [551]. Other principles regarded the careful selection and training of workers depending on the task and a proper division of responsibilities between "*management and the workmen,*" collaborating constructively. He extensively illustrates his scientific management with the operation of pig iron handling. "*The pig-iron handler stoops down, picks up a pig weighing about 92 pounds, walks for a few feet or yards and then drops it on to the ground or upon a pile.*" For this operation of moving 40 kilograms of pig iron at a time, Taylor shows how to improve the daily productivity from 12.5 tons per day to 47.5 tons per day, i.e., to repeat the operation of transporting 40 kilograms of pig iron 1,156 times per day, in such a way that a carefully selected and well-trained workman can continue to perform this task in the long run.

The improvement of the pig iron operations is founded on a design that specifies the "*best way*" of executing these operations. This design was scientific as it was based on detailed time studies of the operations. Taylor made innovative use of the stopwatch for his time studies. More or less concurrently, Frank and Lilian Gilbreth developed motion studies, for which they used cameras, as an addition to the emerging Operations Management methods repertoire. These methods aimed to set process standards (in analogy with standardization of parts) to improve manufacturing operations. The universality was further illustrated by the efficiency improvements obtained by Frank Gilbreth in the pre-industrial operation of bricklaying from 120 bricks per man per hour to 350.

Over time, Frank and Lilian Gilbreth developed a presumably comprehensive set of atomic activities of human operators, called Therbligs [222]. The Therbligs defined an alphabet of symbols that enabled them to explicitly chart any sequence

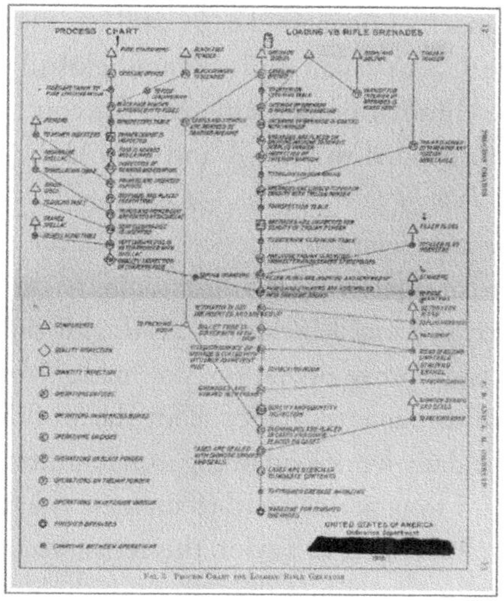

Figure 8.7. Seminal process charts, data source [221].

of human activities. These charts were first used to describe and analyze existing operations. Hence, the set of Therbligs also included inefficient activities such as avoidable delay and rest, which enabled to identify opportunities for process improvement. The Gilbreths also devised "*process charts*" to visually model operations as sequences of activities. In their own words, "*the process chart is a device for visualizing a process as a means of improving it*" [221]. Indeed, process charts were also used to illustrate improved designs, new best ways of working, and new process standards. Figure 8.7 charts the process of loading rifle grenades, one of the two examples in the original 1921 publication [221], in which process charts were presented to a wider audience. It uses standard symbols that are akin to the Therblig symbols, several of which are still common in today's widely used process flow charts.

The operations management practices of the emblematic Ford plant were akin to the aforementioned seminal operations management concepts and techniques. Henry Ford's views included the continuous search for process improvements, for instance, in the form of eliminating waste [202]. Adopting what he referred to as "*the Edison method of trial and error*," he instigated a constant quest for incremental process improvement. At the same time, the quest included to develop discontinuous innovations, replacing manual operations with differently designed machine operations. By 1926, the Ford Company operated around 45,000 machines [202]. This industrialization, which substituted coal, oil, and gas for human energy, enabled Ford to produce far more cars at lower costs with fewer workers earning higher salaries. In subsequent decades, operations managers from Toyota would come to visit and study, and they adopted several of Ford's operations management practices while developing the Toyota Production System (TPS) [272].

The successors of TPS, such as 6σ and "Lean," still use many of the practices that once traveled from the US to Japan and are now globally adopted, such as the process flow charts, the elimination of waste, and the continuous improvement of operating standards (as encoded in PDCA and DMAIC cycles). We revisit these methods in Chapter 9.

The new operations management practices that developed during the second industrial revolution are of interest from a sustainability perspective and, in particular, from the perspective of social inclusion. As Taylor put it, based on personal experience, the efforts to improve operations are made "from the side of management" and often met with fierce resistance from the workforce [551]. There was resistance to increase the daily work load of pig iron handlers almost four-fold from 12.5 tons per day to 47.5 tons per day and to increase the hourly brick laying rate from 120 to 350 bricks. According to Taylor, this resistance could range from informal conversations to unions curtailing the output of their members.

It should be noted, however, that Taylor, the Gilbreths, and Ford claimed that properly designed work and operating standards can serve the interests of employers, employees, and society. Properly designed work is considered to avoid *"hard work"* and *"fatigue"* and to maximize prosperity for employers and employees alike [202, 222, 551]. They advocate that the efficiency improvements come together with a higher wage of 30 to 100 percent, and the latter was realized by Ford in 1914 [202, 551]. Most other manufacturing companies in the Detroit area and elsewhere, however, did not follow this example [605].

Ford's wage increase, together with the Scientific Management principles of collaboration between management and workers and shared responsibility, shows that the operations management quest for efficiency can be socially inclusive. They echo at an organizational level that growth and equity can correlate positively, as described above from an economic perspective. The economic analysis of data over multiple decades and continents, however, reveals that productivity growth and social inclusion have only gone hand in hand under certain conditions [344].

8.7.2 The Continuous Flow Clockwork

As illustrated above, coal, oil, and gas-powered industrial operations changed the scale of operations, numbers, and volumes of production to unprecedented levels throughout the supply chains, from the end consumer products delivered by the assembly lines all the way upstream to the winning of the energy resources at the source of the supply chains. The 2 million standardized black T Fords produced by 1923 implied enormous volumes of materials from suppliers, for instance, including more than 100,000 yards of cotton cloth per day, which in turn triggered cotton and flax production upstream, et cetera [202]. The same applied in other industries, as previously illustrated by the enormous increases in production volumes at sugar processing plants [30]. With all the capital invested in machines, scale increases, and operations management improvements, industrial manufacturing advanced toward low-cost, high-volume production of a continuous flow of products. To this purpose, operations increasingly adopted the principle of interchangeable part, i.e., standardized parts and standardized processes, as introduced in Chapter 7 [618]. Many factory designs were optimized to produce large volumes of a single product without interruption. In the words of Ford, "*no plant is large enough to make two kinds of products.*" These single product layouts to produce a **continuous flow** and high volume of a single product, in the lower right-hand corner of the product variety matrix (see Figure 5.3), became adopted in many industries.

Within such continuous flow operations, human operations were optimized to support low-cost, non-stop production of large volumes by the machines. Many of the tasks of the workers who operated with and between the machines in such operating systems, such as moving materials and loading and unloading of machines, were therefore repetitive and required little skill. Chapter 7 describes how James Hargreaves stood at the origin of this transition toward machine-led, low-skilled work when designing the spinning jenny that enabled his lower-skilled children to be more productive than his highly skilled wife was at cotton spinning.

In his written study of scientifically managing and optimizing pig iron handling operations, Frederick Taylor goes to extremes with low-skilled repetitive operations when optimizing the aforementioned human operation to transport 40 kilogram loads of industrially produced pig iron arriving in large quantities by rail transport and to be delivered to the plant where they are processed 1,156 times per day. Taylor deliberately selects this operation for being "*so crude and elementary in its nature that the writer firmly believes that it would be possible to train an intelligent gorilla so as to become a more efficient pig-iron handler than any man can be.*"

This carefully selected operation shows how men, women, and children were viewed as resources necessary to perform those mechanical and physical operations in large-scale, high-volume machine-dominated settings that were not automated yet. After millions of years of the co-evolution of the physical and mental abilities of humans with their operations, as covered in Chapter 4, the industrial revolution scientifically managed operations to be conducted by humankind in which these abilities were largely disregarded. Humans were managed to operate well below their skill level in an industrial clockwork of machines, as illustrated by Charly Chaplin in Figure 8.8, taken from the movie Modern Times. It has indeed been posited that the invention of the mechanical clock, which forced humans to operate as scheduled by operations managers, has been as important for the industrial revolution as the steam engine (see, e.g., [581]).

Figure 8.8. Modern Times, 1936, data source [105].

In the USA and elsewhere, the continued inflow of immigrants provided a large supply of cheap labor to fill the low-skill jobs that were considered easy to acquire and dismiss, e.g., in response to demand fluctuations [608]. The operations conducted by these workers between the machines were all but disconnected from the natural ecosystems in which most humans worked and lived prior to the industrial revolutions. They worked in an industrial setting and for salary, completing the transition from direct to indirect reward operating models that had started hundreds of thousands of years earlier when humans started manufacturing stone tools and making fire, as described in Chapter 3.

Correspondingly, the discipline of operations management largely developed in disconnect from "the way we live" and the ecosystems in which we live. Operations management is regarded as "the way we work" and is mostly developed in organizations adopting industrial, indirect reward operating models. Operations management is thus focused on organizational goals such as profit, cost, quality, and efficiency. While focusing on these goals, managers and workers were often prone to disregard the effects that the (clockwork of) operations had on ecosystems. The fact that the coal, oil, and gas that powered all the new operations were mined from the ecosystems of planet Earth and that their use affected these ecosystems was hardly a concern at the time.

DOI: 10.1561/9781638282853.ch9

Chapter 9

Communication, Calculation, and All Other Service Operations

The idea behind digital computers may be explained by saying that these machines are intended to carry out any operations which could be done by a human computer.

Alan Turing [564]

9.1 The Operations of Communication

At the start of the 19th century, printed books and written letters were the main forms of information exchange beyond hearing distance. Along the cotton supply chains, from India to Great Britain and to the young nation of the United States of America, information traveled by ship and with the speed of the ship. It could take months or years between sending a letter and receiving a reply. On land, horse speed was the limit for communication beyond hearing distance, e.g., in the form of stage coach networks. In the years 1860–1861, the Pony Express connected California and the West Coast USA with a travel time of 10 days. Until halfway through the 19th century, global empires, multinational enterprises, and large nations relied on wind and animal power for the operations of communication and indeed for the operations management of geographically dispersed organizations and value chains [610].

During the 19th century, however, the technological advancements driving the first two industrial revolutions also affected the operations of producing information goods and services. The operations of communications, for which the preceding advancements have been covered in Chapter 6, changed radically from 1840 onward when the value of newly emerging communication modalities was increasingly recognized and adopted. It marked the start of an information revolution that progressed alongside and beyond the industrial revolutions covered in Chapters 7, 8.

The increase in demand for the raw materials iron and coal that fueled the mass production operations of the industrial revolution implied a need to increase the transported volumes of these raw materials beyond the capacities of horse-pulled trains advancing slowly along the tracks. The steam engine, and more specifically, the steam-powered locomotive, formed a threshold innovation in the industrial revolution that enabled the large-scale transportation of coal and iron required to scale up manufacturing operations [595]. The first steam-powered train started operation in February 1804, connecting the Penydarren Iron Works (furnaces) with the Glamorganshire Canal, both in Wales, from where the iron would be further transported by ship [528, 595].

In subsequent decades, single-track train networks were developed in Great Britain, France, the USA, and several other countries. As the trains on these tracks became heavier and started to travel with greater speed and frequency, it became more important to avoid accidents and, in particular, to avoid the collision of trains approaching one another from opposite sites on the single track. This established a need for communication at speeds beyond the speed of trains pulled by steam-powered locomotives and triggered the adoption and application of information technology telegraphy. Precursors and prototypes of telegraphy were readily available at the time [610]. The first telegraph networks developed along the railroad tracks more or less concurrently around 1840 in France, Great Britain, and the USA [610].

In the following decades, telegraphy would also become an important communication modality for government and business operations beyond railway transportation. As the networks widened, this greatly affected the speed of information exchange over land. By late 1861, the telegraph network of the USA spanned more than 50,000 miles of wire and rendered the Pony Express obsolete [201]. By 1866, telegraph networks had crossed the Channel to connect Great Britain with Europe and the Atlantic Ocean to connect Great Britain with the USA [393]. Within 30 years, the speed of global communication operations had increased almost infinitely.

Train and telegraph networks importantly supported the military operations of the Northern Unionists during the American Civil War and are considered to have determined the outcomes of battles [201]. The telegraph networks would continue to grow in subsequent decades, employing 75,000 operators by 1920 to process 155 million messages annually [406, 516].

From an operational perspective, telegraphy-based communication was not particularly efficient. Every message had to be (dictated and) written down, delivered to the telegraphy office, where it was subsequently coded (e.g., using Morse code), and sent. On the receiving end, it had to be decoded, written, and then physically delivered.

Edison, who had worked as a telegraph boy when he was 16 years old, was one of many engineers working on improvements to telegraphy and, more generally, to the operations of communication. Among many other innovations, he contributed to the development of telephony, a technology in which Alexander Bell and his Bell Telephone Company (later AT&T) would play a leading role [610]. From an operations perspective, telephones were much more efficient than telegraphy, as they eliminated the need for writing, encoding, decoding, and physical delivery steps. The first phones sold in 1877, however, were not suitable to replace telegraphy as they were sold in pairs and the phones of the pair were connected to a direct line. They were mostly sold for professional use [610]. It would not be long, however, before telephones were connected to a switchboard, from where they could be connected to a network of other switchboards and to all phones connected to these switchboards.

It was the job of switchboard operators to connect callers to the respondent of their choice. Following the telegraphy example, the first switch board operators were teenage boys. Their service attitudes and communication skills, however, were considered insufficient for switchboard operations. Some have even been reported to meet customers to "*fight it out*" [57]. Starting in 1878, it became a female profession, employing 3,000 female switchboard operators by 1890 [57, 93]. Bell Telephone Company (later AT&T) held on to these service operations long after dial phones became available to connect the caller automatically with the intended respondent in the late 19th century [119]. The total employment of telephone operators peaked at 357,000 in the USA in 1950—almost double the number of workers at the Ford factories in the hay days of the model T Ford [96, 202]. By then, there were 43 million telephones in the USA, and more than 222 million calls were connected on an average business day [567]. Figure 9.1 depicts a room of female switchboard operators and supervisors.

A new large information service sector had come into existence, telecommunications, with new operations, jobs, and performance to be managed. The expected average call handling time by the operators was 3.5 seconds [93]. In addition to efficiency, service operations management tightly supervised operator behavior, postures, speech, and "*tone of service*," as "*the operator must be a paragon of perfection, a kind of human machine*" [93].

Figure 9.1. Switchboard operators at work, early 20th century, data source [614].

Another device Edison had envisioned to remedy the operational inefficiencies of the telegraph was the electric typewriter. The electric typewriter was designed to type down telegraphed messages at the other end of the telegraph line, thus making human decoding and writing redundant. Edison devised this electric typewriter in 1872, after having been invited to see the first commercially viable typewriter developed by Christopher Sholes [136]. While the adoption of electric typewriters would still take several decades, the mechanical typewriter would soon further propel the information revolution. Mark Twain ventured to write the novel Tom Sawyer on a Remington typewriter in 1874 [300]. From an operations perspective, it is noteworthy that the *QWERTY* keyboard of the Remington typewriters—and subsequently many other typewriters— made it easy for salesmen to type "type writer," yet it was less efficient than alternative keyboard designs in actual typing operations [141].

Typewriters were to be adopted for business purposes far beyond the operations of professional writers. By 1950, there would be over one and a half million stenotypists, typists, and secretaries operating typewriters in the USA, 95 percent of whom were women [96]. By 1970, less than a century since the first typewriter became commercially available, there were more secretaries than farmers in the USA [98, 99]. Their job description typically mentioned to conduct information processing operations such as stenography, typewriting, and telephone communications. It should be added that much of the operations conducted by secretaries in the increasingly large bureaucratic organizations in which they worked have been characterized as "*invisible*" [602].

For the purpose of brevity, we leave many other important and revolutionary advancements in the operations of communication unaddressed, among which are the gramophone, motion pictures, radio, and television.

9.2 The Operations of Calculation

Returning again to the early 19th century, we encounter an era in which Charles Babbage and other mathematicians spent much of their time calculating various tables, among which were nautical tables and logarithmic tables. Charles Babbage considered these mathematical operations to be a work of utter drudgery, involving "intolerable labor and fatiguing monotony" [610]. More than a century and an industrial revolution later, Konrad Zuse echoed these sentiments from his hometown of Berlin when complaining about the "big and awful calculations" his work required him to perform [610]. Zuse would eventually become one of the pioneers to develop a working computer, i.e., a machine that could replace a "human computer" as generally envisioned by Alan Turing in 1950, as quoted in the opening lines of this chapter [564].

For Babbage, and in his time, the words calculator and computer would refer to humans performing mathematical operations, even though he daydreamed about steam-powered machines performing calculations and eventually designed a mechanical calculator [610]. Advancements on this design and related ones would lead to commercially available calculators by the late 19th century. Electronically powered calculators were developed, for instance, to ensure that the processing of the data from the decennial US censuses would not take more than 10 years [271]. They provided automated service operations.

As ballistic warfare advanced in the first half of the 20th century, the calculation of ballistic tables was considered to be of increasing importance for military operations. Moreover, these calculations became more complex and tedious with the development of new ballistic projectiles. Thus, already before Zuse started to develop computers in his parents home in Berlin, the Ballistic Research Laboratory of the USA acquired a "differential analyzer" from MIT professor Vannevar Bush to speed up the mathematical operations required to solve large numbers of higher-order differential equations [87]. The Electronic Numerical Integrator and Computer (ENIAC) would succeed the Differential Analyzer in 1946 [233]. This "*general purpose electronic computing machine*" was able to perform 5,000 basic arithmetic operations (such as addition and subtraction) per second [233]. When presented to the press, the ENIAC developers showed that it would take human computers three days to calculate the "*trajectory of a shell that took 30 seconds to go from the gun to its target*," while the differential analyzer would do it in 30 minutes. The ENIAC calculated this 30-second trajectory in just 20 seconds, "*faster than the shell itself could fly*" [29]. The ENIAC machine capable of performing these operations so quickly weighed 30 tons, existed in 17,000 tubes, occupied more than 200 square meters, and consumed 150 kilowatts of power to complete the task [253].

Another information processing machine that would greatly affect military operations as well as the advancement of information technology was the Enigma. The Enigma was a machine to encrypt information in order to protect the security of information exchange, even if intercepted. With the appearance and some of the functionality of an electronic typewriter, it encoded text entered via the keyboard, which could be decoded by a recipient using another Enigma with the same settings. It was partially developed in Germany and first offered commercially in 1923 [126]. As has been the case with many of the computers developed in the first half of the 20th century, few businesses perceived it to be of value for their operations. Instead, it gained interest from the military in a variety of countries, especially Germany itself. Over its lifetime, more than 100,000 enigmas have been produced and sold [213]. As is well documented, the British government collaborated intensively with the Polish, French, and American governmental agencies in their efforts to break the various and increasingly complex encryptions of the Enigma used by the German armed forces [610]. By 1945, Britain's decoding "*factory*," formally known as the Government Code and Cypher School at Bletchley Park, employed a workforce of 9,000 [126]. The factory used machines called "bombes," which consisted of connected Enigma replicas and were necessary to effectively read the dynamically encrypted Enigma messages of the German armed forces. It is of interest to note that the facility was called a factory even though it produced services rather than goods. Computation and calculation, and other forms of information processing, are service operations, as further discussed and defined below.

The encryption methods of the Enigmas advanced over time, and more advanced machines were needed to decode the messages. Three special-purpose computers were built and put into successful operation by 1943 for this task. Soon after, the more powerful Colossus computers were developed. The second version, of which three machines were built, had 2,400 valves and could process 25,000 characters per second. It was in operation by 1944 and served to decode the highly classified messages encoded by Lorentz SZ40 and SZ42 machines. In retrospect, the decoding operations at Betchley Park are recognized for their significant impact on World War II [126, 213].

Embodying the transition toward the second half of the 20th century, the Colossus and the ENIAC are considered to have been the first electronic general-purpose computers [610]. Progress on the advancement of computers continued after the war, even when the urgency of developing more powerful machines had diminished. Government and academic institutions would form the leading customer segments over the next decade as businesses continued to show limited interest. By 1956, around 200 computers were in operation globally [128, 610].

Various new computers were introduced in the mid-1950s, among which was the IBM 650. According to IBM, the 650 was developed to be operated by "*ordinary businesses,*" and the company delivered almost 2,000 of them until ceasing production in 1962 [281]. This "*workhorse of modern industry*" had the capacity to read 200 punch cards of information per minute and write 100 punch cards of results per minute [281]. It served to speed up engineering calculations, e.g., in the aforementioned ballistic domain, and clerical operations to calculate commissions, process payrolls, make actuarial calculations, prepare customer bills, et cetera [281].

The world hosted more than 25,000 computers in 1964, the year in which the IBM 360 was launched [128]. There was considerable variety in functionality and speed among the IBM 360 models and accompanying products. The fastest models were more than a thousand times faster than the ENIAC and could execute 16 million operations per second. Within the first month of launching the IBM 360, more than 100,000 "*positional orders*" were placed for the various IBM 360 models and accompanying products, of which 7,700 were shipped by the end of 1966 [128].

The IBM 360 triggered wide further adoption of computers in corporations. Computers increasingly affected the operations of hundreds of thousands of businesses, governments and other organizations, and of many millions of employees around the globe. By 1970, the total sales of computer systems and services exceeded 10 billion USD [128]. In the process, the computer industry itself had truly taken off, bringing a variety of new jobs, among which new were professions with new operations such as software architects and programmers. IBM alone employed a workforce of 265,000 by then [128]. Figure 9.2 shows two persons operating an IBM 360.

As already briefly mentioned, valves formed the essential hardware components enabling the operations of the early computers, such as the Colossus and the ENIAC. During the 1950s and 1960s, the valves were increasingly replaced by transistors and subsequently by integrated circuits [610]. The IBM 360 used hybrid integrated circuits. Calculators and computers operating on integrated circuits using silicon as a semiconducting basis were introduced in the late 1960s [610]. As Moore observed as early as 1965 [389], "*Integrated electronics will make electronic techniques more generally available throughout all of society, performing many functions that presently are done inadequately by other techniques or not done at all......For most applications, semiconductor integrated circuits will predominate. Semiconductor devices are the only reasonable candidates presently in existence for the active elements of*

Figure 9.2. Operating an IBM 360, data source [22].

integrated circuits. Passive semiconductor elements look attractive too, because of their potential for low cost and high reliability..." Moore also estimated and predicted the number of components per integrated circuit to double annually [389]. And so it happened indeed, driving the advancement of exponentially faster computers with corresponding increases in memory, of much smaller sizes, and at a lower cost than the IBM 360 and its predecessors.

The first personal computers (PCs) became available in the mid-1970s, targeting enthusiasts and hobbyists. Compared to the preceding business machines, interest grew quickly, and more than 1 million PCs were sold in the year 1980 [476]. In the 1980s, a range of software applications, such as spreadsheets and word processors, became available, which greatly promoted further interest in professional use. By 1990, more than 20 million PCs were sold annually [476]. These annual sales numbers would exceed 100 million by the turn of the millennium and 300 million a decade later, with a value of over 200 billion USD [476, 541]. By then, more than a billion PCs were in operation globally.

Moore's prediction that "*integrated circuits will make electronic techniques more generally available throughout all of society, performing many functions that presently are done inadequately by other techniques or not done at all*" not only regarded the billion personal computers Pcs present on planet Earth some 50 years later, but also the incorporation of integrated circuits in many other personal and household appliances and in manufacturing equipment, ranging from watches, televisions, refrigerators, and cars, to airplanes, computer numerically controlled machines in manufacturing, and automated nutrition systems for animal farming in the agricultural industry [610]. They not only impacted the operations of computers but potentially and increasingly impacted almost any operation conducted or controlled by humankind, whether at work or otherwise. These developments co-evolved with the advancements in communication technology presented below.

9.3 Ubiquitous Information and Communication

The *Compatible Time-Sharing System (CTSS)* was developed in the early 1960s for predecessors of the IBM 360, such as the IBM 704 and 709. The operating system CTSS enabled users, such as programmers, to effectively get their jobs executed between the processing of batch work by the machine [133]. In 1964 (or 1965), three programmers working on the system proposed that "*a new command should be written to allow a user to send a private message to another user which may be delivered at the receiver's convenience*" [135]. This new command went under the name "*MAIL*" and initiated electronic mail.

Electronic mail made a second step when the first wide-area computer network, Arpanet, was established by the end of 1969 (while wide-area experiments date back to 1965) [336, 354]. Email services were developed for Arpanet in 1972 by Ray Tomlinson and installed in the same year [336]. By 1974, three-quarters of the traffic over the 20-node Arpanet consisted of email [354]. While originally developed to advance military operations, Arpanet advanced far beyond this scope when forming the backbone of a newly emerging network called the Internet in 1985 [336]. Five years later, in 1990, the Internet counted more than 2 million users, and soon web browsers and websites developed. At the turn of the millennium, the Internet hosted 413 million users. By 2020, there were more than 4 billion users operating online, and hence, the Internet users outnumbered the non-users [514].

The same US Government Advanced Research Projects Agency that drove the development of Arpanet also drove the development of radio signal-based computer networks in the 1970s [354]. Such advances are parallel to advances in mobile telephone technologies, which date back to at least the 1920s [190]. By 1973, the first fully hand-held mobile phone that connected to the existing wired telephone network became available. Take-up was slow in the years after because of the limited capacity of the required radio frequencies available for mobile phones. Since 1978, cellular networks have been introduced to connect mobile phones, or cellular phones, to nearby antennas serving as network connections within their geographical cell [190]. First-generation mobile phone technology was to be replaced by second-generation (2G) technology in the early 1990s and would be followed up by a third-generation (3G) before the turn of the millennium [190]. These three generations, as well as subsequent 4G and 5G technologies, enabled increasing volumes of information to be communicated, thus increasingly enabling real-time exchange of large data volumes to support operations. The resulting ubiquitous connectivity is further supported by technologies such as WiFi and Bluetooth.

By 2020, close to 3 billion mobile messaging app users communicated through hundreds of billions of messages per day, and the number of emails sent daily was in the same order of magnitude [539, 540].

Naturally, all these advances in information and communication technology often changed operations in ways that rendered existing operating models obsolete. We have already learned that the Pony Express went out of operation when telegraphy developed. Telegraphy, in turn, has long been replaced by telephone and subsequent technologies, causing telegraph boys to lose their jobs. Subsequently, landline telephones have been and are being substituted for mobile phones, and switchboard operations have been abandoned. As some of these developments are recent and ongoing, one might take the viewpoint that the 3rd industrial revolution has not finished yet. More so as all these telecommunication devices apply integrated circuit technologies and more generally build on the ongoing evolution of computer technology. The same applies to other products and devices we have studied in the present and previous chapters, such as clothes, cars, refrigerators, and machines. Even homes, buildings, and roads are being equipped with 3rd industrial revolution technologies, enabling new functionality and operations. Chapters 10 and 11 view these advancements from the perspective of the 4th industrial revolution.

9.4 The Service Revolution

9.4.1 Defining Services

Over time, the agricultural revolution covered in Chapter 5 has largely eliminated pre-existing land-based food procurement operations, more specifically, hunting and gathering. Thus, the importance of the agricultural revolution has been to replace the operations of food procurement through hunting and gathering with food production operations. Not all food procurement operations have been replaced. Hunting and fishing operations are still being practiced in many societies, and the same goes for gathering, e.g., of berries or mushrooms. Some small human subpopulations continue to predominantly rely on hunting and gathering for food procurement until today. By and large, however, the agricultural revolution has irreversibly altered the operations of the primary sector. Likewise, the recent industrial revolutions have also rendered many of the pre-existing manufacturing operations of the secondary (manufacturing) sector obsolete. In fact, the shift from human-powered and performed operations to fossil fuel and electrically powered machine operations has rendered the phrase manu-facturing (making by hand) increasingly inappropriate for the automated production facilities it refers to.

Food production and manufacturing involve the production of tangibles, of goods. We now turn to the development of operations, creating intangibles, of services, which form the tertiary sector. A seminal reference on services is from as recent as 1933 and relates them to a tertiary stage of development that may occur when the first stage of agriculture and the second stage of manufacturing have been mastered [199]. At the same time, it recognizes that services and service operations already existed before the tertiary stage. Service operations may not have left historical traces due to their intangible nature, but they date back to long before the agricultural revolution. Chapter 4, for instance, discusses how service operations such as making fire or keeping a fire burning may have emerged and advanced without leaving evidence millions of years ago. Chapter 5 mentions spiritual and religious services being provided 40,000 years before present. It is also self-evident that hunters and gatherers have provided health services, transportation services, and education services among each other, to mention three of today's largest service sectors.

Thus, the tertiary sector existed long before the agricultural revolution and the industrial revolution. The latter can be quantified if we adopt the common (somewhat imprecise) practice to measure the value created by food production, manufacturing operations, and service operations by the GDP contributions of the primary, secondary, and tertiary sectors, respectively, as applied in national accounting systems. Figure 9.3 shows that the tertiary sector contributed almost 40 percent to GDP in the US and provided more than 20 percent of employment in 1840, the early days of the first two industrial revolutions. Soon after, the tertiary sector would be the largest of the three, and it would only grow as the second industrial revolution unfolded. By 2010, the service sector had more than doubled in its relative contribution to GDP to a dominating 80 percent, and its contribution to employment had almost quadrupled to 83.5 percent. At present, only one in six members of the American workforce works outside of the service sector.

The growth in service operations has not been equally dramatic in all industrialized countries, but many have experienced similar developments, even if somewhat later [262]. In the European Union, the tertiary sector formed 71 percent of GDP by 2005 [184]. The tertiary sector contributed more than half of global GDP in 2018, after experiencing considerable growth in the three preceding decades [546].

Before addressing any questions about service revolutions, it is important to advance the quantitative understanding of Figure 9.3. It relies on definitions in which the primary sector includes agriculture, fishing (a form of hunting), and forestry; the secondary sector includes manufacturing, mining, and construction; and the tertiary sector includes everything else. This remainder consists of sectors such as the government, the financial sector, healthcare, education, retail, transportation, hospitality, and entertainment. It is often asserted that the tertiary sector loosely defines the service sector. The logic of the three-sector theory (see, for instance, [199, 262]) is therefore not based on a formal distinction between

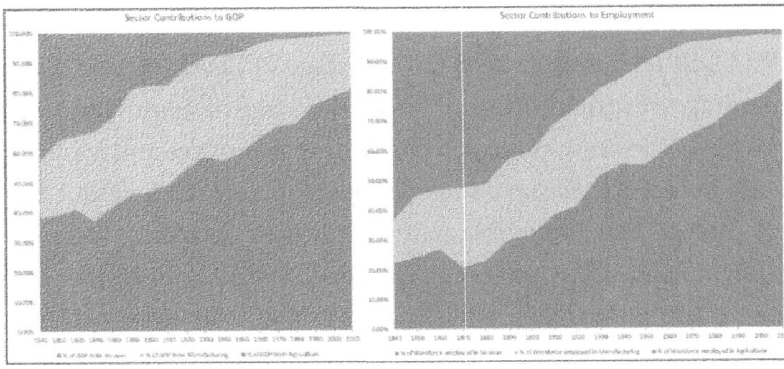

Figure 9.3. Sector contributions to GDP and employment, USA 1840–2010 (19th-century GDP contributions are from preceding year), data source [425].

goods and services, nor does it consider a formal definition of services. Because of our emphasis on operations, however, it will be convenient to adopt a formal distinction between goods and services and hence to define

---

Definition: Service operations are the subset of human operations that produce intangibles.

It is helpful to additionally recognize **blended operations**, which produce a bundle of goods and services.

The value created by service operations as defined above may differ substantially from the value reflected in the contribution to GDP depicted in Figure 9.3. The tertiary sector differs from the service sector if we let the service sector be defined by the collection of all service operations. Many of the operations of organizations categorized to belong to the primary or secondary sector can be viewed as service operations, as further illustrated below. To make things worse, many of the operations of organizations categorized to be part of the tertiary sector produce tangibles and therefore are not service operations. Indeed, the fact that many operations are blended operations inhibits a clear distinction between the sectors on the basis of tangibility.

It has additionally been pointed out that the measurement of the value created by service operations poses a variety of challenges that result in inaccuracies. Some of these challenges relate to the intangible nature of services, which causes their value to be underestimated in national accounts or, more generally, to be inaccurate [243, 508]. It may also be noted that many service operations are not (yet) part of the formal, measured economy but instead are part of the largely unrecorded, informal, or "shadow" economy [513]. Typical service operations of the shadow economy range from self-conducted household operations to babysitting, illegal taxi services through ride-hailing apps, corruption in public services, and drug trafficking. The size of the shadow economy has been estimated to be around ten percent of GDP for advanced economies such as the US economy [512]. The International Labor Organization estimates that more than 60 percent of the global workforce is informally employed [67].

To illustrate the complex relationship between service operations, services, and the three-sectors considered in three sector theory, and hence the measurement of value created by service operations, let us consider the operation of making yogurt from milk, a process that actually predominantly involves operations conducted by bacteria yet involves some human operations. If these operations are conducted on the farm where the cows produce the main raw material, the milk, then yogurt is considered an agricultural product, and the value of the yogurt is fully included in the contribution to GDP of the primary sector. If the farmer instead sells the milk to a diary plant, which then produces the yogurt, the added value created by the diary plant is included in GDP as contributed by the manufacturing sector and hence by the secondary sector. If the milk is bought by a restaurant owner who produces yogurt from it, the added value created by the restaurant goes under the tertiary sector (subsector hospitality) into GDP. Such regardless of whether the yogurt is sold to a dinner having it for dessert at one of the tables of the restaurant (thus enjoying a bundle of the product, yogurt, and the hospitality service) or to a drive-in customer who only buys the yogurt from the take-away window without receiving any services. Lastly, if the yogurt is produced and consumed at home by a person who bought the milk at the farm, the value created by making yogurt at home is not included in GDP but is part of the unrecorded informal economy. All four allocations comply with the United Nations System of National Accounts [150].

9.4.2 Formalization and Servitization

The yogurt example above illustrates two important developments that have contributed to the growth of the tertiary sector without actually impacting service operations: *formalization* and *(de)servitization* [311, 604]. **Formalization** refers to the (mostly) 20th-century development in which many of the operations of life have become operations of work. Humans may have stopped making their own yogurt at home as they lacked time to do so after taking a paid job. Next, they paid someone else to produce and serve yogurt, such as a delivery service, a restaurant, or a person providing household services. More generally, food procurement (delivery services) and preparation (delivery, restaurants, and semi-prepared meals) are prime examples of services that have been formalized. The same holds for child care, elderly care, gardening, cleaning, and many other household services. For household operations, this process is also known as marketization [207]. We may recall from Chapter 3 that even the informal agricultural service operations of bees pollinating flowers (while gatering food) have been formalized by beekeepers that rent out beehives per day to farmers.

The formalization of work has been especially associated with the service sector and has caused service operations to be increasingly represented in GDP. This development has been importantly driven by the increased female participation in formal employment. Many of the new jobs in the service sector were taken by women. Above, we have already seen that new service operations such as switchboard operation and (steno)typing were predominantly performed by women. The informal service operations, such as household operations, these women previously conducted subsequently got formalized as well. The formalization and outsourcing of service operations, such as cleaning, cooking, taking care of the children, et cetera, created new formal jobs in which women were also well represented. In the USA, females made up 31 percent of the workforce by 1940 and 40 percent by 1970, as tertiary sector employment grew from 7 million to 18.2 million [590]. Women took an equal share in absolute formal employment growth over these three decades, as female formal employment almost tripled in absolute terms.

Thus, we may observe that the tertiary sector and GDP grew by recognizing and including the value of already existing yet previously unrecorded service operations. The formalization development therefore relates to the case made in Chapter 2 for a broad definition of operations that includes the operations of life in addition to the operations of work. Operations impact sustainability, regardless of whether they have (yet) been formalized.

The discourse on (de)servitization is wide-ranging and includes a variety of views and definitions [311]. Before diving deeper into its importance for service operations, let us define both servitization and deservitization from an operations perspective.

Definition: Servitization is the modification of operations producing a bundle of outputs consisting of at least one tangible and zero or more intangibles to operations that produce a bundle with the same or less tangibles and more intangibles.

A car company that starts offering an extensive service package (including maintenance and road-side repair) to car buyers practices servitization. A car manufacturer that extends its operations so that its cars can be leased instead of bought also practices servitization.

Definition: Deservitization is the modification of operations producing a bundle of outputs consisting of at least one intangible and zero or more tangibles to operations that produce a bundle with the same or less intangibles and more tangibles.

An example of deservitization arises when the car manufacturer that includes maintenance and road-side repair services in a car sale outsources these operations to subcontractors or deletes one or both from the sale altogether.

When a manufacturing company servitizes heavily, it may become considered a service company rather than a manufacturing company and considered part of the tertiary sector. Alternatively, it may deservitize and outsource some of its service operations (such as maintenance) to a service company. Both of these examples of servitization and deservitization cause the tertiary sector to grow and the secondary sector to shrink, while the same operations remain being conducted and GDP is hardly affected.

Early examples of servitization dating back to the late 1960s, for example, regard selling airplane engines per service hour (flight hour)—instead of selling the tangible engine—and selling printed sheets rather than printer machines [311]. These early examples of servitization show that operations—whether conducted by machines such as airplane engines or printing machines, by humans, or otherwise (e.g., pollination by bees)—are intrinsically intangible and hence can be offered as a service. Not only the airplane engine can and has been servitized, but service companies can provide a full flight as a service to an airline company, including the plane and the crew. Production operations can also be provided as a service, and car assembly is being provided as a service [311]. Does that make car assembly a service operation, and should its added value be attributed to the tertiary sector and considered evidence of economic development?

9.4.3 More Old Service Operations, or Revolutionary New Service Operations?

The growth of the service sector cannot be fully attributed to formalizing or (de)servitizing existing "old" operations, and we explore drivers of service sector growth in depth in this section, linking it to the advancements in information and communication technology where applicable. We conclude by shedding light on the subsection title: has there been a threshold development resulting in revolutionary new service operations, or have we rather been able to automate and scale up old services?

The hierarchy of needs hypothesis posits that demand for many services increases as income increases with GDP. At lower levels, much of the available income is spent on basic needs such as food and housing, which mainly involve products from the primary and secondary sectors. Thus, the hypothesis suggests that demand for services corresponding to higher needs increases more than proportionally with increases in GDP [508]. The sustained increases in GPD are, for instance, associated with increasing demand for health services, education, hospitality, and entertainment. It has been estimated that health services constituted around 2 percent of GDP by 1900 in the US, after which it grew, partially by formalization, to 13.3 percent by the year 2000 and exceeded 17 percent by 2010 [156]. By then, more than 10 percent of the workforce was employed in health service operations [320]. In the same year, 2010, the 10 percent of the total workforce employed in the leisure and hospitality sector outnumbered the number of people working in manufacturing in the US [320]. Increases in consumer demand for services have been estimated to account (roughly) for one quarter of the growth of the service sector (see [508] and the references therein).

Increases in the demand for business services make a comparable contribution to the growth in service operations [508]. The growth in business services can, for an important part, be explained by the absolute growth in agriculture and manufacturing. As absolute primary and secondary sector outputs increased so impressively during the industrial revolutions, the need for complementary business services such as transportation services, financial services, and so on increased as well [293]. Moreover, the growth of these business service companies generated further demand, for instance, when financial institutions require information and communication services.

With these developments in mind, it then becomes relevant to observe that for some operations, human labor is "*a requisite for the final product*," whereas in other operations, human "*labor is itself the end*" [47]. The labor productivity of services that require human labor for final value delivery has often not been substantially impacted by the technological advancements made since the onset of the industrial revolutions. Hence, the human labor intensity of such service operations can explain how the combination of GDP growth and increases in labor productivity in the primary and secondary sectors, in which technology replaced human labor together, have enabled the growth of employment in the tertiary sector, where such substitution was less feasible [47]. Thus, increased demand for such labor-intensive "old services" has driven growth in value creation and employment in the service sector.

Let us now explore potential "new service" and how service operations have been impacted by the 3rd industrial revolution. This revolution has delivered machines and tools that have automated many information and communication services, thus substituting labor for capital and increasing labor productivity in service operations. The opening lines of this chapter explicitly refer to such a substitution when suggesting that "*machines are intended to carry out any operations which could be done by a human computer.*" These improvements have yielded new services, such as long-distance communication. As the 3rd industrial revolution progressed, it increasingly enabled automation of old service operations for which human labor had previously been supposed to be an end, such as the replacement of tour guides by audio devices and apps [48].

If a switchboard operator handles one call per 3.5 seconds, more calls mean more switchboard operations, whose number grew to 357,000. Telephone services, however, became much less labor-intensive when connections were established (faster and) automatically rather than as the result of a service operation conducted by a courteous human switchboard operator. Such technological advances have indeed changed many service operations at their core, improving their value and increasing labor productivity (often reducing costs as well) [562]. It has become possible for millions of consumers to watch movies at home, each selecting to their individual taste among a huge number of alternatives and at their own time of choice. If well equipped, they can watch it on a large screen with advanced sound effects and pause and resume at any time as desired. Contrast this with a cinema theater in which several hundred people can watch a movie played from tape by an operator, without sound, except for live piano, twice a week. Or with the ancient Greeks who went to see a play by Sophocles in an amphitheater?

Despite the technological advancements, however, the play still needs actors, and the human labor and service operations of acting have not yet seen revolutionary productivity improvements as a result of the 3rd industrial revolution. For such service operations in which some human labor remains essential, any scale-up resulting from labor productivity improvement (and cost reduction) in some of the service operations can then grow employment for the operations that remain labor intensive (and for which there is no productivity or cost improvement) [48]. Indeed, the movie industry employs many actors and creates a considerable part of the value included in the tertiary sector. It has been estimated that the growth in such human labor-intensive service operations, in which productivity improvement has been (partially) stagnant, accounts for around half of the service sector growth [508].

From a service operations perspective, the advancements are not so much in the increase in the number of actors. The revolutionary developments are the use of film on tape to screen a play in a cinema rather than play it in the theater, and the use of digital technology to enable watching it at home. These new information and communication technologies have caused entirely new service delivery operations, some of which were subsequently abandoned because newer technological advancements brought a next generation of service operations that made the previous ones obsolete. Like the operator in the cinema, other jobs conducting new sets of operations emerged and disappeared as technology advanced, such as telegraph boy, switchboard operator, typist, et cetera. Other new jobs designed to perform new sets of service operations that emerged through the 3rd industrial revolution and which have so far survived are, for instance, camera (wo)man, sound engineer, programmer, web designer, influencer, et cetera.

All of these new jobs are evidence of new service operations that are essentially different from "old" service operations that have been automated. It would be hard to find pre-industrial revolution equivalents of switchboard operators, sound engineers, or programmers. Likewise, the 3rd industrial revolution has resulted in devices conducting service operations that are not automated versions of service operations conducted by humans before it started, such as telecommunication satellites, pacemakers, or search engines.

The advances in the tertiary sector, resulting in jobs such as switchboard operator, IBM 360 programmer, and web designer, further disconnected the operations of work from the ecosystems of planet Earth, as discussed in Chapter 8 in relation to the secondary sector. Moreover, the increasing scale and automation resulting from the 3rd industrial revolution also further enlarged the distance between the operators in the primary and secondary sectors to the ecosystems (see, e.g., [204]). As a matter of course, many may have remained unaware of any effects of their operations on the ecosystems of planet Earth and on the sustainability of these ecosystems.

9.5 Sustainability of the 3rd Industrial Revolution

There is no clear-cut starting date for the 3rd industrial revolution from which to start measuring its impact on sustainability. Did it start with telegraphy, with telephony, with the first mainframe computers, or did the revolutionary part only take off in the 1980s when PCs and the Internet took shape? Moreover, and closely related, it would be helpful to have a date at which the 2nd industrial revolution ended. So far, there are no commonly agreed-upon dates for either. Even if we define the 2nd industrial revolution to relate to operations for the production of goods and the 3rd industrial revolution to relate to operations for service production, there is much overlap in time and in operations considered as both of these types of operations are closely related. Wireless Internet services require many hardware components. Hence, the measurement of the separate sustainability effects of the 3rd industrial revolution is far from straightforward, if well defined at all.

9.5.1 Economic Development

As before, let us first view the economic developments in the US, where many of the developments took place early, have remained relatively advanced, and are well studied. We recall from Figure 9.3 that the service sector has grown to form 80 percent of GDP and hence has become a main driver of any economic development. Before examining it more closely, it is worth establishing that this growth has been enabled by and interacted with developments in the primary and secondary sectors.

Over the period 1840–2010, the total workforce in agriculture started out at 3.57 million and increased until the early 20th century, after which it diminished to 2.23 million. The relative contribution of the primary sector to GDP and employment decreased much more dramatically over the same period, respectively, from 42.6 percent to 1.1 percent and from 63 percent to 1.7 percent. As the national real GDP grew by a factor of 264 over the same period [65], these numbers imply a remarkable increase in agricultural labor productivity by a factor of 10.9. In simple words, much fewer humans conducting agricultural operations produced much more agricultural output and provided food for a much larger US population in 2010 than in 1840. We thus may take the view that the increases in agricultural productivity have freed up human resources to work in the operations of other sectors, and in particular in the service sector. More so, the labor productivity growth in the agricultural sector over this period was higher than in manufacturing, which went through two industrial revolutions (resulting in a labor productivity growth of 10.4) and in the rapidly growing service sector (with a labor productivity growth of 6.5). We refer to Chapter 5 and [371] for further reading on the recent "agricultural revolutions" that drove the agricultural productivity growth.

For the secondary sector, which includes the manufacturing industry that hosted two industrial revolutions within the time frame of Figure 9.3, we can observe that it has grown from a contribution of slightly below 20 percent of GDP to around 35 percent by 1940, after which it diminished again to around 20 percent by 2010. Comparable secondary sector trends are commonly found in developed countries, many of which peaked later and below 40 percent of GDP [199, 262]. More generally, evidence suggests that countries that developed their manufacturing sectors later peaked at lower levels [176]. It has already been discussed above how relative and absolute employment in manufacturing has diminished, in part because machines continue to replace the human workforce.

The service revolution and industrial revolution have not only built on the huge advancements in agricultural operations but have also enabled them. The cotton gin and sugar refinery processes discussed in Chapters 5, 8 increased the value created by agricultural production. Contemporary agricultural operations use machines powered by combustion engines, such as tractors, and information technology to optimize crop and livestock production [132].

The service sector has also enabled productivity growth in manufacturing. Services can contribute to manufacturing as inputs (for instance, when workers are hired through employment agencies), support manufacturing operations (as is the case for plant maintenance services), and add value after end product delivery (after

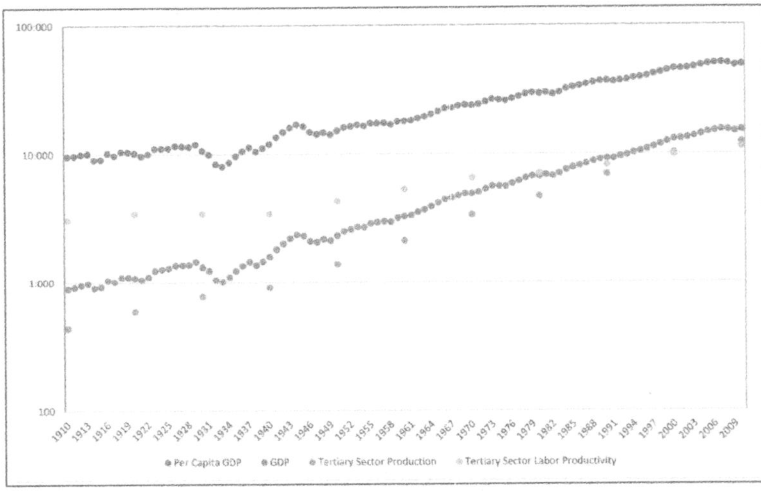

Figure 9.4. Per capita GDP growth and tertiary sector labor productivity in the USA in 1910–2010, as well as GDP and tertiary sector production. Amounts in 2011 USD (billions for GDP and tertiary sector production), data source [425].

sales services). Thus, the three sectors and their operations are interrelated in a variety of ways, with agricultural operations forming the basis and service operations building on the other two [427].

In the US, the relative increase of the tertiary sector became more significant around 1910 (see 9.3), and Figure 9.4 depicts a century of GDP and GDP per capita growth from 1910 onward on a logarithmic scale, accompanied by the corresponding service sector growth. The logarithmic scale shows exponential growth and, at the same time, makes it clear that the tertiary sector has grown more quickly than GDP.

The developments of global (per capita) GDP starting from 1920 (the latest year covered in the previous chapter) to 2010 are depicted in Figure 9.5. Global GDP data for the tertiary (or service) sector development over most of this period are not available. However, using data from multiple sources and countries, Eichengreen and Gupta find evidence of a positive nonlinear association between service sector growth and GDP [176]. The service sector contribution to GDP appears to grow in two waves, of which the first wave is mostly related to traditional services that predate the 3rd industrial revolution, whereas the second wave is driven by new information- and communication-based service operations. The value created by these new service operations caused the service sector to grow more rapidly than GDP in the second stage.

While the above findings broadly align with difficulties experienced to see advancements in information and communication products and services reflected in GDP growth for other economies and at other time periods, it is noteworthy that

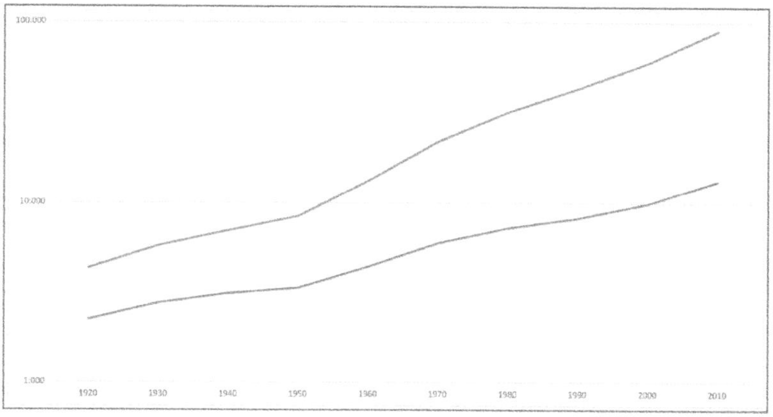

Figure 9.5. Global per capita GDP growth, 1910–2010. Amounts in 2011 USD (billions for GDP), data source [425].

IT investment has been significantly related to GDP growth for developed countries (OECD) but that this relationship has been insignificant in the 20th century for developing countries, perhaps because requirements such as a corresponding communication infrastructure or a matching workforce were lacking [450, 530].

The growth of the relative contribution of between-country inequalities to global inequality, which commenced in the early 19th century and has been driven up by three successive industrial revolutions, peaked between 1980 and 2000 (depending on data and measures used), when it contributed well above half of the global inequality [103, 382]. By then, GDP growth in countries that industrialized early on and had developed large service sectors, such as in North America and Europe, started to level off, while an increasing number of other countries industrialized and automated their operations, resulting in higher GDP growth rates. The global between- country income inequality has decreased relatively quickly since peaking, and by 2020 it was back at the levels of the heydays of the 2nd industrial revolution, roughly a century earlier. Global inequality itself also peaked toward the end of the 20th century and has also declined considerably since, to the levels of the heydays of the 2nd industrial revolution and before [103, 382].

From the above, it appears that the most recent 3rd industrial revolution has reduced inequalities and eventually favored developing countries. This can be partially explained by the globalizing effect the information and communication technology has had on economies. The advances in information processing and communication have importantly enabled agricultural and manufacturing supply chains to be managed effectively over longer distances. They have facilitated the coordination of source-to-sink value chain operations while partitioning them into increasingly smaller subsets of operations and stretching over larger parts of the planet Earth [20].

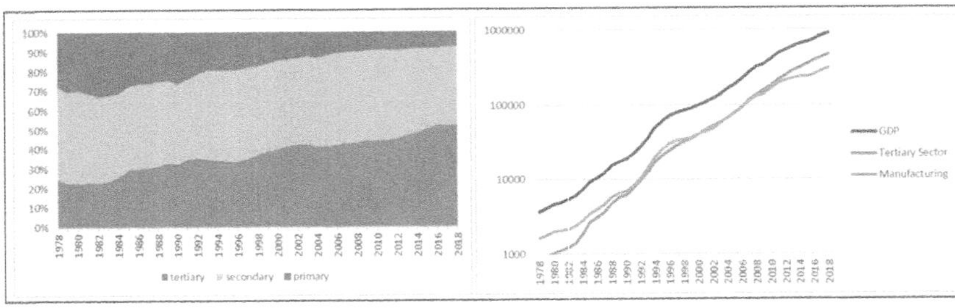

Figure 9.6. China: Sector contributions to GDP and GDP growth, 1978–2018. Amounts in 100 million 2019 Yuan, data source [631].

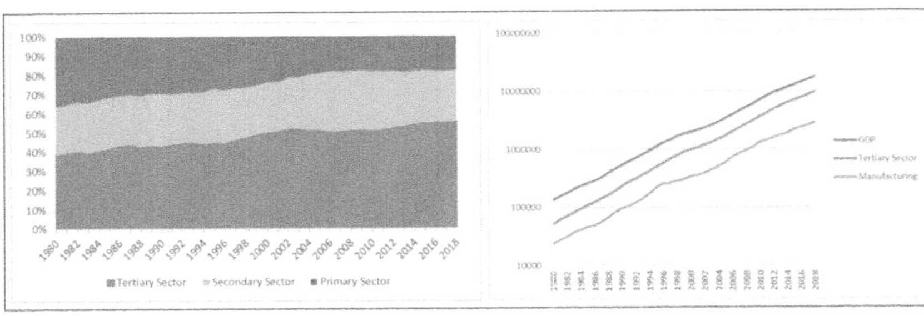

Figure 9.7. India: Sector contributions to GDP and GDP growth, 1980–2018. Amounts in 1,000 2021 crore, data source [149].

For China, the resulting opportunities have translated into a quickly growing manufacturing sector (which formed a relatively large part of GDP) by providing manufacturing operations to supply global demand and thus producing well beyond domestic demand [631]. Figure 9.6 shows that manufacturing contributed around 40 percent of GDP for most of the years since the 1978 reforms until 2011, during which GDP grew rapidly and exponentially. In 2018, more than half of the manufactured goods were exported and accounted for more than 90 percent of China's exports. The three main categories of exported goods were all related to the 3rd industrial revolution: automatic data processing machines, telecommunication machines and equipment, and electronic machines and parts. Until 2015, the secondary sector as a whole was the largest sector, and manufacturing operations alone contributed more to GDP than the tertiary sector as recently as 2006. Since 2015, the operations of the tertiary sector have contributed most to China's GDP.

Figure 9.7 shows that throughout the same time period, the tertiary sector contributed most to GDP in India, growing toward 60 percent of GDP. Around 35 percent of the Indian workforce was employed in service operations by 2018 [149]. Until 2010, the majority of the Indian workforce was employed in agricultural operations, their number being relatively stable in absolute terms but decreasing quickly in relative terms. Manufacturing quite stably contributed around 17 percent to GDP over the almost four decades considered in Figure 9.7. For India, service exports have helped to drive an exponential GDP growth that is comparable to the GDP growth of China [149]. India has shown exceptional growth in service exports, which formed more than half of all exports by 2015 [175, 314]. This growth was especially driven by growth in modern information and communication technology-based service operations such as software services and call center services [175]. While India's modern service export growth is exceptional, it reflects the global development in which growth in trade in modern services outgrows goods trade [175, 385].

Taken together, China and India illustrate how manufacturing operations and service operations in global value chains can drive rapid economic growth. The management of these global value chains essentially depends on information and communication technology advances. Moreover, many of these value chains produce information and communication goods and services. Through their global coverage, their effects on GDP growth are extensive and have resulted in reductions in income inequalities between countries. So far, however, these reductions have only partially undone the growth in inequalities produced since the onset of the first industrial revolution. Moreover, the recent improvements in value generated by the operations in China and India have been accompanied by large increases in within-country income and wealth inequalities, which have reached historically high levels [103]. These increases are representative of a global trend in within-country economic inequality that has accompanied economic growth in the first decades of the third millennium.

9.5.2 Social Inclusion

In Sections 9.1, 9.4, we have learned that women were well represented in the employment for many of the new jobs that emerged as a result of the 3rd industrial revolution. Switchboard operator and typist served as typical examples of jobs. Hence, these technological advances resulted in including more women in formal operations. Moreover, the formal employment of women implied they engaged less in informal activities such as housekeeping and created a formal demand for these service operations. This, in turn, created jobs in which women were often well represented. The growth in other service sectors such as healthcare, hospitality, and education also entailed employment opportunities for women. Altogether, the growth of the service sector has resulted in including more women in formalized employment and, hence, in a more balanced gender division of formal work and pay. Such as the disappearance of former (in)formal employment as servants by private households.

Another, parallel, avenue toward more inclusive operations provided by the 3rd industrial revolution is the demand it has generated for a more highly skilled workforce. While the initial 2nd industrial revolution brought jobs in manufacturing that were very physical (and some were even said to be performed more efficiently by great apes [551]), the new service jobs emphasized cognitive and social skills rather than physical ability. Together with the automation of purely physical jobs in manufacturing, this has made education and knowledge acquisition more valuable. Conversely, the increased labor productivity of a better-skilled workforce made human labor more valuable and facilitated higher salaries (for those skills for which supply did not exceed demand on the labor market). Thus, as we have seen in Chapter 8, the 20th century has known a long period of more inclusive welfare, first within developed countries and later at a global level. Within countries, however, these developments have been reversed roughly since the turn of the century. The increases in value created by society through the information and communication technologies of the 3rd industrial revolution may have favored correspondingly skilled and resourced subpopulations in comparison to subpopulations that are less resourced, less skilled, or skilled otherwise, and thus have enlarged welfare gaps.

9.5.3 Environmental Protection

For many decades, the information and communication sector has duly followed Moore's law, formulated in 1965, of continued exponential growth of computer speed and corresponding operational use [362, 389]. Indeed, computer sales have grown to several hundreds of millions annually, and annual sales of smart phones have risen to above 1 billion [476, 541]. The use of these hardware devices has contributed to even more rapidly growing data volumes exchanged over wireless networks and the Internet. Hundreds of billions of emails and social media messages are being exchanged annually. These communication operations have, in turn, grown the number and size of data centers [56, 362].

In view of the above, the expectations have been that the production of the hardware and the subsequent energy use while in operation would cause a proportionally increasing carbon footprint. Interestingly, this has not been the case. The growth in hardware has been compensated by efficiency improvements, and the same holds for the energy use of the devices. Moreover, the contribution of renewable energy to the energy used by data centers, network operations, and the like has grown [56, 362]. Altogether, it has been estimated that the global carbon footprint of the operations of the information and communication sector is not growing and that the same applies to the entertainment and media sector (which also includes paper-based products) [362]. Moreover, the total contributions of these sectors are in the order of 3 percent of global GHG emissions [56, 362].

While the sectors most closely related to the 3rd industrial revolution play only a modest role in global GHG emissions, these emissions have grown almost ten-fold

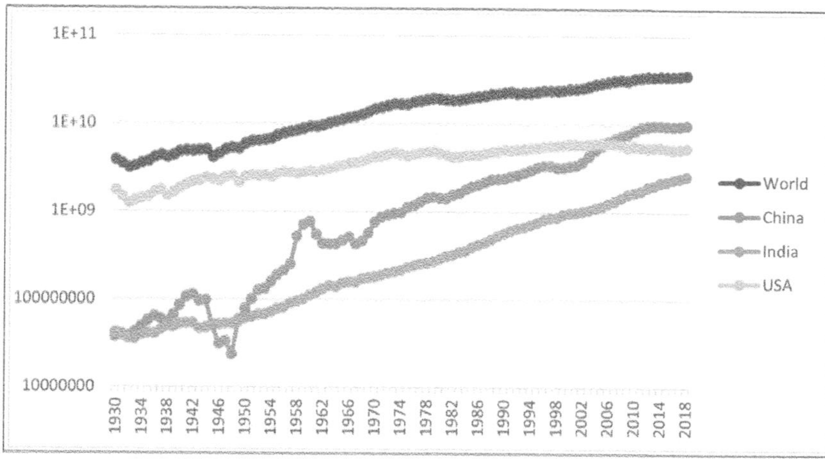

Figure 9.8. Global GHG emissions for China, India, USA, and the world since 1930 (in tons of CO_2e) [623].

since 1930, as depicted in Figure 9.8. They have tripled in the USA with its rapidly developing service sector, after leveling off by the end of the 1990s. In India, where the service sector has been such an important component of economic growth and manufacturing plays a relatively modest part in the economy, GHG emissions have grown by a factor of more than 60 over the period 1930–2018. In China, where manufacturing is important for economic growth, GHG emissions have grown by a factor of 260. It may be noted, however, that while China has surpassed the USA in total GHG emissions, its per capita GHG emissions are still much lower than in the USA, as holds even more true for India.

Hence, we may ask whether the tremendous growth in GHG emissions over the last century has mostly resulted from more and more energy-intensive manufacturing operations, or whether, for example, service operations and the operations of life of humankind also have significantly contributed to this growth, despite the modest role played by the operations of the information and communications sector? To answer this question, let us recall from Table 1.1 that the primary and secondary sectors together account for roughly two-thirds of global GHG emissions (when including direct use by the energy sector). The remainder is divided over buildings (17.1 percent) and transportation (14.7 percent). Transportation includes the transportation of goods in supply chains as well as transportation for travel services (e.g., holidays and leisure activities), restaurant delivery services, and for the operation of life such as commuting and going shopping. Additionally, almost two-thirds of building-related emissions are from residential buildings. Thus, even when allowing a margin to accommodate the unknown yet significant contribution made by informal service operations, the service sector, including the transport sector, is unlikely to have contributed more than 20 percent of total GHG emissions in 2018.

The numbers above, and particularly the major contributions to GHG emissions by the agricultural, manufacturing, and energy sectors, still play an important role for service operations and the operations of life. More so as global value chains lead toward end consumers and their operations of life. Agricultural produce ultimately contributes to consumption at home (whether as food for humans, clothes, or otherwise). Likewise, all manufactured goods directly or indirectly contribute to the production of goods and services consumed during the operations of life. The energy value chains also feed into all stages of these value chains, whether agricultural, manufacturing, services, at home, or otherwise. The large service sector emits GHGs from the energy required for service operations. These emissions are not straightforward to establish, as we have seen that many products can be easily servitized. The GHG emissions of an hour of air travel or assembly line operations are independent of whether the airplane (or assembly line) is bought, leased, or purchased as a service. This servitization perspective shows how the carbon footprint of service operations includes the carbon footprints of procured products required to deliver the services.

Following the value chain perspective, we may allocate all the GHG emissions to end products and services purchased by end customers for their operations of life. For each end consumer, this sums up her or his carbon footprint. This perspective will be further elaborated in the next chapter when reviewing current operations and their sustainability. It is already worth mentioning that the framing of value chains in terms of producing, value-adding, and steps that lead toward a final consumer is challenged by circular economy principles, in which processes are viewed as cycles (circles) rather than chains [191, 299].

The production of information and communication technology devices impacts the sustainability of the ecosystem of planet Earth not only through GHG emissions. These devices are composed of a variety of hazardous materials, among which are hazardous metals and plastics. Thus, the production of more than 1 billion mobile phones annually, added to hundreds of millions of PCs, routers, TV screens, et cetera, causes environmental pollution throughout the value chain, starting from the mining of these materials to the manufacturing, transport, et cetera. Moreover, the use of these devices is often far from circular, and the devices typically end up as e-waste after a lifetime of less than a decade [7]. This even holds true for metals such as lithium used in rechargeable batteries, which improve sustainability by replacing single-use batteries.

The presence of production spills and electronic waste in the form of end-of-life devices and their components pollutes ecosystems and threatens the health and lives of species living in these ecosystems. Electronic waste, for example, in the form of lead, lithium, or PVCs, can contaminate the air, soil, and water and has been associated with an increased prevalence of forms of cancer, diseases of the nervous system, reproduction and development disorders, cardiovascular diseases, kidney and liver diseases, et cetera [7]. Unfortunately, the locations in which electronic waste is processed, for circular reuse purposes, have been shown to be particularly hazardous to the (often informal) workers and their livelihoods [7, 328]. Many low- and middle-income countries are disproportionally affected by pollution, such as that caused by producing and recycling electronic devices and batteries [328].

9.6 Operations Management Perspectives

The development of the telegraph, the typewriter, and subsequent information and communication technologies have profoundly altered how humankind manages the operations of work and life. In fact, operations management has been a prime purpose for many of these developments to have taken place. Telegraphy was first developed with the purpose of managing transportation service operations. Email programs were first developed as messaging systems to coordinate programming operations. The same holds, in part or in full, for many other advancements of the 3rd industrial revolution.

Below, we elaborate on two relevant contemporary perspectives on the main interactions between the 3rd industrial revolution and operations management. First, we consider the direct contributions of information and communication products and services to operations management in the primary and secondary sectors. We contrast advanced operations management technologies with less technology-intensive alternatives, in particular Lean Manufacturing and take a view toward their sustainability. Second, we look at service operations management practices, highlighting developments in three of the most relevant service industries, hospital and leisure, healthcare, and information and communication services.

9.6.1 The Role of Information and Communication Technology in Operations Management

Toward the end of the 19th century, the increasing speed of transportation, together with telegraphy, and telephony induced more frequent ordering of smaller quantities [609]. The scientific management principles of Taylor (see Chapter 8) included methods for the very careful planning of all resulting production activities, including the ordering of raw materials and all successive processing steps. Following scientific management methods, the Production Office was in charge of all these preparations. This office was designed to use an extensive, card-based, information system to manage materials and production. Despite its rigor and potential, the uptake of Taylor's tedious system remained very limited.

As the material and resource requirements of ordered end products in large manufacturing plants became increasingly complex and difficult to manage, the need for a system to manage these operations became more pressing. The information processing tasks of Taylor's system were automated step by step and started to be executed by accounting machines in the 1930s. These automated systems gradually developed into integrated information systems over the next decades, using the rudimentary programming and computing functionality of the time [609]. In the early 1970s, IBM developed a standardized information product, a software system for the IBM 360 named COPICS, which incorporated the *Materials Requirements Planning (MRP)* methods [285, 423, 424].

COPICS caused MRP implementation to spread much more rapidly, and other MRP software providers would join the market [285, 423, 424]. These MRP software systems advanced from Materials Requirements Planning to Manufacturing Resource Planning (MRP) and then to Enterprise Resource Planning, as currently widely implemented in manufacturing industries and in many service industries [285]. These systems not only support the primary business operations, such as a manufacturing process, but also the operations of secondary business functions ranging from purchasing to finance to human resources, et cetera [285]. Contemporary ERP systems are accessible as software products and as services.

Together with the increase in digital communication capacity, ERP software and other software supporting business operations enabled the integration of primary and secondary business operations over multiple functional departments, over multiple sites, and indeed over multiple links with subsets of operations in the value chain. These information technologies thus enabled the management of the continuously complex and dispersed global value chains [20], including the electronics supply chains that provided the necessary hardware. This hardware started to include other technologies providing supply chain management information, such as barcode scanners, radio frequency IDs (RFIDs), RFID tags, and readers. The planning systems in turn were integrated with computerized manufacturing systems, including manufacturing robots, as further highlighted in Chapters 10 and 11.

The MRP logic is based on planned demand and creates schedules and material flows to meet planned demand. The schedules and flows are explicitly revised as time progresses and planned demand changes. The computerized MRP systems are capable of doing these planning revisions quickly and accurately, even for operations involving multiple end products composed of hundreds of components or more. MRP "engines" are capable of maintaining the feasibility and timeliness of production operations. The MRP logic does not necessarily yield the most efficient production plans. One type of solution for the problem of finding an optimal production plan has been the development of "*Advanced Planning and*

Scheduling (APS)" software, which uses more advanced optimization algorithms from the Operations Research domain to plan and schedule operations.

The uptake of information and communication technology to manage operations has also impacted the primary sector in which agricultural value chains are rooted and from which they emanate across the globe. While not all implementations are equally beneficial, a wide variety of improvements in cost reduction and the value of produce have been reported, ranging from the use of mobile applications informing smallholder farmers in Tanzania about weather forecasts and market prices to the application of smart aerial vehicles in Europe to manage weeds and minimize the use of pesticides [31, 49, 464]. Information technology also plays an increasingly important role in managing industrial forms of livestock production, such as fish farming and meat and dairy production [204, 287].

9.6.2 Lean Operations and the Elimination of Waste

As we have learned in Chapter 8, operations managers from Toyota have visited US manufacturing companies in the early days of MRP and adopted several of the operations management practices encountered while developing the TPS [272]. This inspired them to develop Lean Manufacturing methods, which are based on a set of principles that are quite different from the MRP principles. Lean Manufacturing is led by actual demand and directs production operations in the supply chain to replenish inventory reductions caused by actual demand.

Tajichi Ohno laid the foundations of TPS, believing that "*much of the excess information provided by computers is not needed for production at all*" [418]. The information technology used to manage operations supplying components to the assembly line used (paper) cards and vinyl envelopes that were attached to containers (kanbans) as late as the final decades of the 20th century. The "(kanban) cards" create a "demand pull" mechanism as the kanban card taken from an empty parts container serves as a production order to replenish one container of the same parts. This simple card- and envelope-based information technology is reminiscent of the token-based inventory management systems that preceded the development of the script (see Chapter 6). Over time, electronic kanban systems have been developed as well, and kanban systems have been incorporated into ERP systems [458].

Kanban systems enable just-in-time material flow and have been co-evalled with a continuous emphasis on the elimination of waste and the prevention of quality problems. The waste elimination focus of Toyota was one of the learnings from their visits to American manufacturing. We have seen in Chapter 8 that Taylor and Ford also continuously pursued waste elimination [202, 551]. TPS much more systematically and comprehensively identified and eliminated seven types of waste and, for instance, considered stock on hand to be wasteful [418]. In the supply chain feeding into the assembly line, the management efforts were therefore devoted to ensuring that materials required for an operation arrived "*just in time*" [418].

Working just in time, i.e., with little or no stock in hand (as enforced by allowing a limited number of kanbans), causes problems whenever machines are defective or components produced fail to meet standards. Hence, quality management and the elimination of the root causes of quality issues have been essential to successful Lean Manufacturing. The principle of "*autonomation, or automation with a human touch,*" has enabled the management of continuous quality improvement [418]. The emphasis on the prevention of defective production is built into (automated) machines, and operators take responsibility for eliminating the root causes of any quality problem, thus preventing their recurrence. Compared to the times and thoughts of Taylor and Ford, Toyota did not view the operators as a physical resource between the machines but rather as a person taking shared responsibility for production operations and for the problem solving required for the never-ending improvement of operations.

The success of Toyota's operations management methods (and of akin operations management practices developed in post-World War II manufacturing in Japan) is evidenced by the decline of the American car manufacturing industry since the 1970s. The top three American car brands (Chrysler (now Fiat Chrysler), Ford, and GM) held over 80 percent of the US market in 1975, and Japanese car brands were well below 10 percent together. By 2010, the market share of the American top three had almost halved, and the Japanese top three (Honda, Nissan, and Toyota) held more than one-third of the US market [304].

The principles developed by Toyota and other Japanese manufacturers have developed further, and their success has led to their dissemination across the globe and across many industries. The emphasis on the elimination of waste, i.e., of non-value-added activities, has become known as "Lean Manufacturing," and the never-ending quest for quality improvement is presently mostly known under the name 6σ. The phrase 6σ refers to extremely low probabilities of operations not producing according to specifications [396]. As we shall see below, lean principles and quality management have also spread to service operations management across a variety of service industries. All these lean adoptions share an emphasis on avoiding of waste, i.e., operations that fail to add value for the end customer. Indeed, lean views operations as wasteful if they are not adding value in terms of the price the end customer is willing to pay [418].

The definition of waste considered in lean is consumption-oriented and not based on environmental waste. In view of our research aims, one may subsequently wonder how lean relates to sustainability. Does the elimination of waste promote sustainability?

On the positive side, the elimination of waste typically results in discontinuing subsets of operations that are not adding value to the products and services delivered to the customer. Ceasing to perform such wasteful operations may then free resources to contribute to economic development otherwise. Moreover, the eliminated operations no longer contribute to GHG emissions, no longer require or produce toxic and hazardous materials, nor harm ecosystems otherwise. Thus, elimination of waste can indeed quite directly improve sustainability [448].

On the other hand, lean can negatively impact sustainability. What if end customers attach no value to paying decent wages for working schedules following commonly agreed standards? What if end customers attach no value to halting child labor, equal pay, environmental protection, preserving biodiversity, reducing GHG emissions, et cetera? In such cases, paying decent salaries can be viewed as a waste, and hiring cheaper child labor may appear to eliminate the waste of paying unnecessarily high salaries. Such lean practices may not be implemented in transparency and escape the awareness of end customers, even if they value sustainability, as the operations may take place far from the continent on which they live.

9.6.3 Service Operations Management

Service operations have been practiced since long before the agricultural revolution, and service operations management therefore necessarily dates back to the early stages of the development of humankind. Chapters 4 and 5 mention "old services" such as trading, transportation, and religion, which date back to before the agricultural revolution, and the development of more recent "old services" such as governmental administration services and military services in the civilizations that emerged as a result of the agricultural revolution. Chapters 7 and 8 witness how important service sectors such as the transportation sector and the retail sector have developed alongside manufacturing during the first two industrial revolutions in the same global value chains. Banking and financial services form another subset of service operations that gained importance as industrial operations became more capital-intensive.

Inevitably, the management of service operations dates back as long as these services themselves, and service operations management has therefore not emerged in the 20th century nor has it evolved as part of the 3rd industrial revolution. Still, service operations management is often viewed as a relatively young branch of operations management that has emerged well after the contributions of Frederick Taylor and his contemporaries to manufacturing operations management.

Perhaps the rifle loading process flow chart developed by Gilbreth (see Chapter 8) can be viewed as one of the first contributions to service operations management from this era in which operations management gained recognition as a management discipline. Indirectly, this process flow chart already emphasized virtually eternal service quality dimensions such as timeliness and compliance with standards. "Quality" itself was first defined and discussed more than 2,000 years ago by Plato, who related it to tangibles and intangibles alike [570].

The global and steady growth of the service sectors that started in the 19th century has also brought a wide range of "new services," all of which were affected by the 3rd industrial revolution. Let us reflect on these developments by highlighting some of the service sectors that have advanced most significantly. For instance, the health services sector warrants further analysis as total healthcare expenditure amounted to 3.8 trillion USD in 2019, well above the total value added by manufacturing of 2.4 trillion USD [155, 170].

Leisure and Hospitality Service Operations Management

As mentioned earlier in this chapter, the leisure and hospitality services sector provided more employment than the manufacturing sector in the US by 2010, which was far from the case before the start of the 3rd industrial revolution. This development relates to a more general theme regarding the amount of time spent on the operations of work versus the time spent on the operations of life. This topic has been of interest throughout the timeline of human operations, starting with hunting and gathering, and continues to be of interest in relation to the upcoming 4IR. We therefore now first study leisure and hospitality services, how they expanded, and what new, distinctive, operations management perspectives they have brought along.

It is quite likely that, like present-day hunter-gatherers, humans enjoyed much more leisure time before the agricultural revolution than most of humankind has enjoyed since. Hunter-gatherers likely ended their hunting and gathering operations when their daily (direct return) subsistence needs were met [42, 581], after which they had time to turn to the operations of life. In civilizations that have emerged since the agricultural revolution, working hours tended to increase, even though leisure time continued to be highly valued. The ancient Greek society held little esteem for most operations of work and highly valued a wide variety of cultural activities, ranging from theater plays to Olympic games. The Romans organized their "circenses," which are still symbols of their well-established leisure services sector.

In more recent times, the leisure services sector gained momentum in industrialized countries as the long working hours of the first industrial revolution started to decrease in the mid-19th century and welfare increased, elevating the importance of higher-order needs. This held particularly true for the "*leisure class*" that was increasingly populous by the end of the 19th century and whose members were "*free of industrial occupations*" and had much time for the operations of life [581, 582].

Figure 9.9. Food assembly in McDonald's first kitchen. Photo: Courtesy of Michael Corenblith.

Cesar Ritz and August Escoffier designed and operated luxury hotels for the leisure class around the turn of the century [183]. The kitchen design adopted process layouts, recipes were standardized for consistency and quality, and included standardized components such as basic sauces [183]. At the other end of the restaurant industry spectrum, the McDonald brothers, who founded the first fast food restaurant of the same name, would go as far as designing a hamburger assembly line for their kitchen almost half a century later, as illustrated in Figure 9.9.

Cesar Ritz meanwhile developed innovative and highest standards for hotelling and hospitality, including the introduction of a private bathroom for each hotel room [183]. Over the course of the 20th century, the Ritz Carlton hotels would expand on the quality philosophy of Cesar Ritz and set uncompromisingly high service standards, referred to as the "*Gold Standard*" [274]. By the end of the century, it had thoroughly implemented total quality management principles and adopted many of the quality management principles originally developed by Toyota. The quality management practices, for instance, included a process flow chart for the standard operating procedure to be followed when meeting a guest [205]. At the same time, they highly relied on empowering all employees to operate in compliance with the highest service standards and "*fulfil even the unexpressed wishes and needs of our guests*" [205].

An important service operations management principle developed by Cesar Ritz is that "*The customer is always right.*" While this statement is unlikely to be a universal truth, it does highlight that service quality is perceived individually and is difficult to assess objectively. Customers make their own subjective quality assessment, which importantly determines the customer value provided and the customer satisfaction [432]. Service quality management thus needs to address subjective customer valuations of characteristics of intangibles that are much harder than the often objectively measurable characteristics of tangible goods. Service operations management has therefore advanced quality measurement and management practices beyond those already available in the manufacturing industry, going as far as systematically anticipating unexpressed wishes and needs.

The qualities intended to be provided in the hospitality and leisure industry often relate to hedonic values of pleasure and well-being [280]. Disney can be viewed to have led the way as the Disneyland service operations aimed to provide the "*happiest place in the world*" [574]. These aspirations confirm that the leisure and hospitality services sector has grown in developed countries to fulfill needs that are well above the lower two levels of Maslow's hierarchy of needs [365].

Globally, travel and tourism services had grown to provide more than 10 percent of GDP and jobs and 20 percent in job growth before the start of the COVID-19 pandemic [561]. It is one of the service sectors that provides opportunities for socially inclusive economic growth, especially for women, and for at least developing countries [35, 461]. Tourism accounted for 8 percent of global GHG emissions in 2013, and this percentage was expected to grow with industry growth (despite decarbonization efforts) [338]. If it were a country, it would rank 3rd for GHG emissions, after China and the USA. Further growth of the sector is also associated with other planetary boundaries such as biodiversity, biochemical flows, water, and novel entities [461]. Thus, operations management for this service subsector will need to resolve the incompatibilities of current tourism with social and planetary boundaries, for instance, adopting the RISA framework presented in Chapter 10 [264].

Health Service Operations Management

Whereas the (hedonic) value of the hospitality and leisure services covered above primarily lies in the service experience (and consumed during service delivery), the value of health services is often determined by the resulting health outcomes (obtained after the service delivery). Corresponding service outcomes-based paradigms are elaborated in scientific and policy frameworks adopted across the globe in measures such as health-related quality of life (QALYs) and disability-adjusted life years (DALYs) [505]. Whereas the QALYs and DALYs achieved for human populations and individuals have determinants beyond health services (such as socioeconomic status and lifestyle), health service outcomes importantly influence the health and well-being of humans. For instance, surgery can prevent the disability of blindness caused by cataracts, a condition from which more than 10 million humans suffer globally, and the subsequent loss of health and quality of life [188].

The relationship between a specific health service (e.g., cataract surgery) and health-related quality of life is not direct but mediated via the clinical outcomes achieved by health services (e.g., restoring visual function). Thus, the primary health services operations are the clinical operations conducted by medical doctors, nurses, and other health professionals. The word surgery is derived from the ancient Greek χαιρουργία (kheirourgia), which is composed of χέρι (hand) and δουλειά (work). Globally, surgery services are known as operations and are conducted in operating rooms and operating theaters. Obviously, many health service operations other than surgery exist, as there are many health conditions for which surgery is not (or only a part of) the treatment that forms the preferred health service to be provided.

In part, the growth of the health service sector can be explained by increases in welfare and the availability of financial resources to procure health services, whether through personal or public budgets. This mechanism should, however, not be confused with the tendency of more resourceful humans to satisfy needs that are ranked higher in Maslov's hierarchy, as health is a basic need and equal and timely access to affordable health services is a human right [24, 365]. The growth in health service operations reflects the ability to treat conditions for which no treatment existed previously and the ability to provide better treatments. The health service value chains became more valuable and effective because of the supply of newly developed equipment, pharmaceuticals, knowledge, and other inputs that enabled more effective clinical operations.

Alongside the industrial revolutions, advances in chemistry have been foundational to the development of drugs and treatments and have made a major contribution to new and more effective health services, for instance through the development of antibiotics and chemotherapy [164]. Industrial machines and products have also brought major service improvements, for instance, in the form of X-ray machines and pacemakers. Present-day intensive care units are packed with medical technology, and the same holds for commonly accessible health services such as dental service clinics. All these advancements have increased the volume and complexity of services provided and caused a more highly skilled and larger workforce to be required to deliver a wider range of services.

The health services sector has also grown because of formalization. In many countries, the necessary care for persons with health demands, such as elderly persons or disabled persons, has transitioned from informal care services provided by family members to formal health service provisioning by skilled professionals. In many countries with advanced health systems, the cost of the health workforce presently forms the majority of the operating costs of the health service sector and exceeds 5 percent of GDP [210, 542].

Governmental and professional organizations often play leading roles in developing standards for the clinical operations of health services. The responsibilities for the subsequent operational management are then often allocated among senior professionals, such as the chief nurse or the chair of the medical department. Over time, these clinical operating standards have increasingly become evidence-based, i.e., their effectiveness has been established by rigorously designed scientific research (such as randomized control trials). Such standards relate not only to preferred medication or surgical treatments but also to surgery checklists and hand washing protocols and are increasingly encoded in health information technology (for instance, computerized drug prescription support systems) [361, 501, 506].

Management of compliance with the standards, which includes creating conditions and systems that promote adherence to the standards, can help prevent avoidable errors by professionals provisioning clinical health service operations. Unfortunately, such errors are not rare and were, for instance, estimated to have caused between 44,000 and 98,000 hospital deaths a year in the USA by the end of the 20th century [159].

Errors can be viewed as a form of non-value-added waste created in health service operations, and as was the case for the elimination of waste by Ford and Toyota, the prevention of errors can often lead to lower costs while improving the quality and safety of care in terms of better health outcomes. This is firmly illustrated by the Aravind Eye Hospitals, which, in the first decade of the 21st century, provided cataract surgery in India at 2 percent of the typical cost in the US and with lower adverse outcomes rates [570]. Interestingly, the clinical operations of Aravind Eye Hospitals were inspired by the operational service efficiency of McDonald's (see also Figure 9.10. The low cost and high volume (which has grown to over 300,000 cataract surgeries per year) enabled India to make significant inroads into the burden of cataract disease among its population, specifically the poor [273, 470].

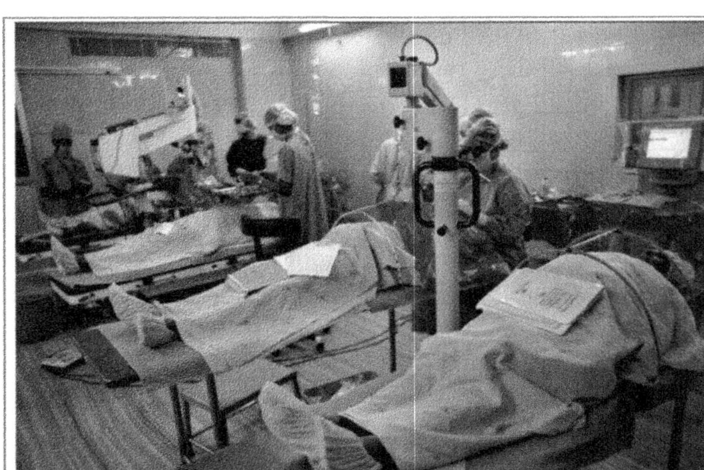

Figure 9.10. High-volume cataract surgery services at Aravind Eye Hospital [379].

From an operations management perspective, it is worth noting that Aravind firmly and explicitly designed its clinical operations based on service operations management practices developed in other (service) industries. This is not common practice in health service operations. Rather, professional knowledge and perspectives often play a leading role in the design, control, and improvement of clinical service operations. This follows the institutionalized logic that professional autonomy and authority are leading in the management of clinical service operations and that organizational management hierarchies beyond the professional domain may have limited influence. The responsibilities of operations managers often explicitly exclude clinical operations proper and rather focus on the coordination and facilitation of the clinical operations, for instance, by ensuring the timely availability of all required equipment and materials.

This somewhat confusing labeling of roles and responsibilities is not unique to healthcare. It commonly arises in service organizations in which highly skilled professionals deliver the main services to customers. In such organizations, operations managers, and indeed management in general, typically hold little operational authority over the highly skilled professionals, whether they are musicians in a symphony orchestra (or a hard rock band), professional soccer players, lawyers, cooks, or scientific researchers. Professional independence, autonomy, and confidentiality can even be formalized in law and regulation, as is the case for the professional autonomy of medical professionals in health service operations. Their legally and professionally encoded responsibilities, governed by the national authorities that issue the professional licenses, go beyond corporate service operations management approaches, even if codified by pledges, credos, and gold standards, as was the case for Ritz Carlton.

In connection with the leisure and hospitality industries, it may, however, be noted that—despite often being considered of lesser importance than the health service outcomes—the service experience of the patient has increasingly received attention in health service operations management. Without going so far as perceiving that the customer is always right, taking the patient experience as the leading quality perspective and adopting a patient-centered approach to the quality of the health service operations provided has become an important standard in recent decades [127]. This has stimulated the adoption of concepts from service operations management, for instance, by considering how health service operations were to be managed "*if Disney ran your hospital*" [333]. Likewise, operations management approaches that are led by customer quality and value, such as Lean Manufacturing and 6σ, have been widely adopted in health services organizations and value chains, be it with mixed results [18, 137].

Sustainability has been a topic of interest for the health service industry for multiple years and is increasingly receiving attention in health services operations management. Initially, the sustainability focus mostly regarded the sustainability of financial and human resources in health service systems at a societal level [197, 419]. Effective health service provision is positively associated with economic growth [63]. However, there are affordability limits to the financial and human resources a society can sustainably allocate to health service provisioning.

Aravind Eye Hospitals presents one of many examples illustrating that health service operations management can contribute to the affordability and cost-effectiveness of health services and thus promote economic growth. As Aravind's operational efficiency enables it to provide free services for the poor, it also illustrates how health service operations management can contribute to economic growth for the poor and to social inclusion.

It is well known that social, economic, and demographic factors such as age, gender, race, education, and income cause differences in health, access to health services, and quality of health services received, both within countries and between countries [73]. The vast majority of countries have significantly increased health service access and quality in recent decades. From a global perspective, however, differences in access and quality have enlarged [40]. These differences reveal improvement opportunities, as also considered by the (sub)goals of the United Nation's Sustainable Development Goal 3 ("*Ensure healthy lives and promote well-being for all at all ages*"), that relate to universal access for all at all ages and specifically emphasize strengthening of health service capacity in developing countries [398].

Health service organizations have recently started to pay more attention to environmental protection [422]. Initial international studies, for instance, show that the carbon footprints of health service operations in OECD countries on average were around 5 percent of national carbon footprints in 2014, and roughly the same applies to the global carbon footprint of health services [337, 447, 481]. If health services were a country, they would have ranked 4th in 2015 for GHG emissions, just after India. The global footprint of healthcare for other air pollutants such as particulate matter and SO_2 is lower but comparably significant [337].

Health services and their supply chains produce a variety of materials, other than GHGs and air pollutants that harm the environment. These materials are often categorized as waste and range from infectious materials that may carry bacteria and viruses, used and unused drug residuals, toxic chemicals, radioactive waste, disposables, et cetera [106]. Moreover, health services provider organizations produce the typical waste of hospitality service organizations, and sometimes in larger volumes (for instance, to comply with standards of hygiene). Clearly, health services operations management, which is so naturally focused on SDG 3, "*Good health and well being for all,*" will have to increasingly consider other SDGs, among which those related to planetary boundaries, such as SDGs 6, 7, 13, 14, and 15 [169].

9.6.4 Operations Management for Information and Communication Products and Services

1964 has been an important year for software development operations. As covered above, it was the year in which electronic mail was developed as an information service supporting software developers to manage and coordinate their operations and in which the IBM 360 was launched [128, 135]. The software development of the IBM 360, together with the hardware development, can be characterized as an unprecedented achievement in information technology development.

After having led much of the IBM 360 software development, Frederick Brooks estimated that, depending on the complexity of the programming task, the yearly production of final delivered assembler language code was between 600 and 3,000 debugged instructions per man per year [79]. This translates to 3 to 15 lines per working day. In addition, he points out that if the size and complexity of programming project tasks have been optimistically underestimated, it often is infeasible to make up for the additional time required by involving more workers, more "*manpower*": "*like dousing a fire with gasoline, this makes matters worse, much worse. More fire requires more gasoline, and thus begins a regenerative cycle which ends in disaster*" [79].

The paradoxical consequence of reduced total productivity from adding manpower has multiple causes. For instance, newly added programmers or testers need to be informed and introduced before being productive, and this requires time from manpower already on task, reducing their productivity while the new ones are still unproductive. Moreover, dividing the software development task among more people means that more communication is required, leaving less productive time and more room for coordination errors [79].

For many decades, the management of software development operations has remained challenging, and the common narrative has remained that software development projects are commonly delivered late, have exceeded their budget, and fail to meet customer expectations. These difficulties have persisted despite the development of project management methods, such as PERT/CPM, and despite adherence to rigorous methods to conduct each of the software engineering phases, such as the phases of the waterfall model analysis, design, program, test, and maintenance [445].

Some have argued that these difficulties have persisted because of the rigorous and detailed planning and management methods that are ill-suited to cope with changes [445]. Many software systems are developed in complex and dynamic contexts that are difficult to capture fully and objectively. It may even be inherently impossible to correctly envision the requirements and operation of ambitious new information systems within such contexts [400, 563].

The above observations caused an interest in software development operations management practices that refrained from adding detail, rigor, and control to remedy the cost, time, and functionality problems. Prototyping, or rapid prototyping, and other methodologies to develop software iteratively and incrementally, based on initial versions shared with users, were the common denominator of the first "*revolution*" in software development operations management [400]. Prototyping methods were already well known in other engineering disciplines.

While scientific evidence of the superiority of new software development management approaches is non-conclusive [91, 386], practice has advanced toward a further embrace of customer proximity and responsiveness to changing needs and circumstances. Further new software engineering methods emerged, such as agile, scrum, and lean software development [51, 91, 453]. Lean software development indeed adopts the lean methods originating from Japanese car manufacturing and has been adopted in the management of service operations in other service sectors as well, as narrated above. Lean software development identifies seven software development wastes that are inspired by the seven wastes initially identified by Toyota [453]. For example, the waste of "fixing defects" translates to the software development waste of "bug fixing." Lean software development has also adopted the use of Kanban boards to implement a customer-driven pull system for software development, which is essentially different from the push-based methodologies of the waterfall model and its predecessors [453].

If all goes well, developed software goes into operation as a component of information systems and enters new stages of its life cycle. These information systems may in turn support other services and goods-creating operations embedded in value chains delivering products and services to final consumers [468]. The MRP and ERP systems discussed above are prime examples of such systems. The apps end customers have on their phones to order food and other tangible products for their operations of life are as well. The operation of such information systems also requires management, and this area of operations management has matured as the 3rd industrial revolution advanced [377, 468].

Lastly, let us consider the growth of value chains delivering information products and services as their primary products and services. The operations of information product and service delivery have advanced tremendously since the time of Bell's switchboard operators, as considered earlier in this chapter. The Internet has become a main delivery channel for information products and services. At present, one can consume the service of watching a movie without visiting a cinema or having an antenna or cable to watch it on television. Movies are being streamed via mobile internet and can be started, watched, halted, and resumed virtually anywhere and at any time. The management of these service delivery operations has opened up a new area for operations management.

An even more novel area of information production and service operations management regards the domain of information products that have developed through the Internet. Among these new services are search engines, platforms to match supply and demand, social media, et cetera. How to operate a search engine? How to operate a ride-hailing application? Operating models for such new services are in the early stages of development and appear to have made only initial connections with the existing body of operations management knowledge. More so as many of these services are provided for free, and the operating models of the companies providing these services require additional revenue generation services to be commercially sustainable. Commonly practiced additional services provided are targeted advertising and access to bundled customer data, both delivered to paying business customers.

The sustainability issues that need to be managed as this young service sector advances have in part been addressed above and will be readdressed in Chapter 11 in relation to the 4th industrial revolution.

DOI: 10.1561/9781638282853.ch10

Chapter 10

A Revolutionary Transgression of Planetary Boundaries

We have two lives and the second one begins when we realize we have only one.

Confucius

10.1 Evolution and Revolutions Toward Contemporary Operations

Our first research aim has been to analyze the history or operations of humankind and their impact on sustainability. The previous chapters have extensively covered the timeline and sustainability of human operations and largely clarify how these have accumulated in present-day operations, operations management practices, and the associated sustainability problems.

After this subsection summarizes and synthesizes the history of operations, this chapter turns to our second research aim and takes inventory of the present urgent sustainability challenges and the ongoing technological innovations of the 4IR. This chapter thus extends the preceding chapters to present-day human operations and operations management practices. Moreover, it serves as a point of reflection and inflection, in the spirit of the proposition by Confucius that opens this chapter.

By providing a structured overview of today's challenges and opportunities, this chapter lays the foundation for a solution-oriented approach toward future, sustainable operations, and operations management in the final Chapter 11.

Today's human operations are a result of 4 billion years or more of the development of operations in the ecosystems of planet Earth, as covered in the preceding chapters [407]. As a starting point for reflection on today's human operations, Figure 10.1 synthesizes these developments. It synthesizes the evolution of operations in four stages, each of which can be viewed to have been triggered by a revolutionary development. For each of these stages, Figure 10.1 subsequently characterizes the main developments in operations and adds several relevant perspectives as they have been systematically elaborated in the preceding chapters: the energy perspective, the intangibles and information perspective, the operations management perspective, and the sustainability perspective.

The first stage covers the development of ecosystems on planet Earth since the emergence of life on earth until the emergence of humans, as highlighted in Chapter 3. The revolutionary transition defining this stage is the entrance of living species into the ecosystems of planet Earth. This revolution marks the point in time before which there was no life on planet Earth, and hence there were no operations, and after which life and operations existed on the planet.

While the emergence of life on planet Earth and its ecosystems can certainly be classified as a revolutionary development, scientific efforts directed at understanding the emergence of life typically view this transition as an evolutionary process that took place at a molecular level over several stages and a long period of time [535].

Several threshold steps in the evolution of the ecosystems of planet Earth occurred after the emergence of life and operations on the planet. Among these are the appearance of the first animals and the appearance of the first terrestrial ecosystems, as covered in Chapter 3. The nonhuman living species populating the ecosystems of the planet in this first stage practiced increasingly advanced operations for which they communicated and used tools and adopted operations management practices such as the division of labor. Moreover, their operations include ecosystem engineering practices, such as the building of dams. Chapter 3 illustrates these developments when describing the operations of bees and beavers.

Most of the species that once lived on planet Earth have become extinct. Typically, extinction has been the result of external disturbances to the ecosystems in which they lived. The remains of members of extinct and extant species populating planet Earth for several billions of years before the emergence of humans partially transformed into high-caloric elements of the ecosystems of planet Earth, such as coal, oil, and gas.

Revolution	Ecosystem & Tangibles	Energy	Intangibles & Information	Operations Management	Sustainability
Life on Planet Earth	Development of pre-human ecosystems. Entrance of living species in ecosystems. First marine species, than terrestrial and aerial.	Carbon production surpluses create coal, oil, and gas reserves.	Pre-human forms of communication (e.g., pheromones, waggle dancing, by bees, birdsong. Genetically acquired knowledge.	Division of work. Mostly direct return.	New species continue to develop externally caused mass extinctions
Bipedal Dexterity	Hominins started operating on the ground walking on their feet and using their hands for other operations. Increased practicing of scavenging and hunting. Hand-made (manu-factured) stone tools, later assembled tools. Humans move to top of food chain entering all continents and many of their ecosystems.	Hand-made fire improved human energy efficiency for food procurement (tools) and digestion (cooking).	Brain size and cognitive abilities increased. Development of speech. Development of spirituality. Learned cognitive abilities, rock art.	Conscious advances toward indirect return and division of labor. Very limited assets/property beyond tools/food in hand. Very limited inventory. No meaningful separation between operations of work and life.	Population growth from thousands to millions. Social inclusive group values and sharing. First steps in human ecosystem engineering. Extinctions of large animals and of other homo species.
Farming & Domestication	Domestication of plants, animals, and humankind. Large-scale ecosystem engineering. Increasingly effective food production, resulting in surpluses, trade. Development of other (domestic) manufacturing and service operations and urban environments.	First machines, driven by human, animal, water, or wind power. Wood burning as a source of power (e.g., ovens for metal working, cooking,...).	Development of the script and information management systems for inventory/trade/operations/property. Advancement of science and knowledge as a resource. Development of book printing, increasing production factor knowledge.	Specialization and formalization of work, professions. Ownership of land, capital goods, human resources, inventories. Global trade/transport networks. Strong organizational hierarchies. Use of information systems. Emergence of completely indirect return operations. Separation of work and life, especially in urban environments.	Economic growth mostly because of population growth from millions to hundreds of millions. Social exclusion. Slavery, coerced labor-based operating models. Large scale ecosystem engineering, e.g., causing deforestation. Unsustainable mining operations.
Fossil Fueled Machining	Development of mass production of goods and services. Substitution of human manual labor by fossil fuel-powered machines.	Use of coal, oil, gas for combustion engines and for production of electricity.	Development of machines for computation and communication, including networks. Digitization of information. Substitution of human cognitive.	Birth of operations management as a recognized organizational discipline. Organizing men and/in function of machines. Elimination of waste. Development of information systems to support management (or manage).	Population growth from hundreds of millions to billions. Per capita GDP growth by several orders of magnitude. Rise of inequity among stakeholders in (global) value chains.

Figure 10.1 Synthesis of the timeline of operations.

The second stage identified in Figure 10.1 starts around 6 million years ago and is marked by the transition of some greater ape subspecies to adopt the practice of walking on two legs when on the ground: bipedal motion. The transition progressed from bipedal motion toward living on the ground. It forms another transition that can be viewed as a revolutionary development that changed operations on planet Earth forever. It marks the entrance of hominin species, some of which later developed into human species, practicing human operations as defined in 2.

Again, this revolution is commonly viewed as an evolutionary process that likely occurred at different locations at different moments in time in Africa, as Savannas increasingly replaced forests because of climate change over a period of millions of years [142, 218, 560]. Bipedal motion may have developed as a Darwinian response to changes in the ecosystem.

The adaptation in operations of these first hominins toward working and living on the ground has been a precursor to several subsequent threshold steps in human evolution, such as their uptake of tool manufacturing operations while developing into increasingly effective hunters. They advanced the production of tools well beyond the practices of other species, enabling their population to move up in the food chain and the growth of body and brain sizes. While Homo sapiens spread across all continents of planet Earth, their operational effectiveness caused the extinction of several other living species. Another threshold development in human operations has been the acquisition of the operational capability to make fire. It enabled the use of other ecosystem elements, such as wood, as an energy resource.

The extinctions of large animals caused by the effective hunting operations of humans promoted the development of alternative food procurement operations. Aided by changes toward a stable, warmer climate, humankind started experimenting with agriculture around 20,000 years ago. After several thousands of years, these developments led to the adoption and spread of agricultural operations to replace hunting and gathering and sedentary ways of working and living. Together with their animals and plants, humankind domesticated and switched largely to live in niches constructed by humankind instead of operating in the (non-engineered) ecosystems planet Earth provided. Moreover, they introduced forms of ownership for their constructed niches. Humans started to consider themselves to be owners of the (parts of an) ecosystem in which they operated. The ecosystem, land, and almost everything on and below the surface of the land became owned resources and capital goods in service of human operations.

This revolutionary transition, known as the agricultural revolution, marks the transition to the third stage in Figure 10.1. It caused tremendous increases in the effectiveness of the food production operations of the primary sector, which in turn enabled considerable growth in the secondary and tertiary sectors. In all of these sectors, there were important subsequent innovations in tools and machines, and hence in the operations, which fundamentally altered the way humans lived, worked, and related to one another.

Among the advancements in operations that followed the agricultural revolution are the development of the script and metal mining and metal production operations. Metal mining is the earliest form of the use of ecosystem resources that are not renewable at timescales meaningful for humankind as a raw material for human operations. Subsequent advances in operations include the invention of tools and machines such as ploughs, wheels, boats, windmills, and—much later—the printing press. Advancements such as the invention of the wheel and the printing press by themselves have fundamentally changed the ways humankind works and lives (see Chapters 5, 6).

Chapter 7 describes how steam engines were developed around 2,000 years before present, and over the past 1600 years, evolved into the machines that powered the onset of the industrial revolution. The switch to widespread use of coal as a source of energy was partially driven by necessity, as the use of wood caused large-scale deforestation. Once again, a change in ecosystems that resulted from human operations co-initiated a change in the operating model that would play an important role in the next revolutionary development.

The switch from wood to coal, and subsequently to oil and gas, also as a resource to produce electricity, characterizes the revolution toward the next industrial stage. This fourth stage in Figure 10.1 relies on the use of ecosystem resources that are not renewable at timescales meaningful for humankind to provide the energy required for human operations. The combustion of these fossil fuels, which are available as resources in the ecosystems of planet Earth, caused and causes climate change and many other sustainability challenges in the environmental protection domain.

The industrial revolutions covered in Chapters 7, 8, and 9 are characterized not only by the use of fossil fuels but also by the machines for which the fossil fuels form the energy required to operate. These machines have replaced and superseded physical and cognitive human labor. As elaborated in these preceding chapters, machines can spin cotton, produce electricity, form an assembly line that produces cars, decipher encrypted information, and exchange emails and other messages across planet Earth. The initial industrial revolution has brought on many subsequent changes in operations that have been characterized as revolutionary and have fundamentally altered the way humans live, work, and relate to one another today, including the car, the computer, and the Internet.

From the above, it becomes clear that human ecosystem engineering has repeatedly caused ecosystem changes that render their operating models unsustainable. Humankind has ran out of large game to hunt and of wood to combust. Exogenous ecosystem changes have also repeatedly played a role, rendering human operations unsustainable or creating circumstances that triggered revolutionary developments in operations. Given the importance of the ecological dimension for the sustainability of human operations, we first examine current sustainability challenges from the environmental ecosystem perspective below and subsequently involve the other two Paris Agreement domains of social inclusion and economic development.

10.2 Transgressing the Planetary Boundaries

10.2.1 Environmental Protection

Adopting the planetary boundaries framework introduced in Chapter 2, Figure 10.2 summarizes the present scientific understanding regarding the nine boundaries identified in the framework [322, 443, 487, 543]. Figure 10.2 shows that human operations have almost surely already transgressed the global boundary regarding biochemical phosphor and nitrogen flows and the boundary for novel entities [443, 543]. Moreover, human operations might have transgressed an additional four of the nine boundaries, which are the boundaries for climate change, biointegrity, land-system change, and atmospheric aerosol loading. For these four dimensions, the direction of change is mostly away from safety and toward transgressing the global boundary, and some of the boundaries are already transgressed locally. The latter applies, for instance, to freshwater use and extinction rates. For the three remaining boundaries, the boundary is most likely not transgressed [322, 543].

Planetary Boundaries and Trangression Measures				
Boundary	Safely within Boundary	Transgression Uncertain	Boundary Transgressed	Measure
Climate Change		■		Atmospheric CO_2 concentration
Biosphere Integrity		■		Extinction rate
Land-system change		■		Area of forested land as percentage of original forest cover
Freshwater use	■			Consumptive blue water use per year
Biochemical flows			■	Global and local P (phosphor) and N (nitrogen) flows
Ocean acidification	■			Carbonate ion concentration
Atmospheric aerosol loading		■		Aerosol optical depth
Stratospheric Ozone depletion	■			Stratosopheric O_3 concentration
Novel Entities			■	Concentration relative to no effect concentration

Figure 10.2. Planetary ecosystem boundary transgressions caused by present global operations.

Let us elaborate on the relationship between today's human operations and two of the most critical boundaries more closely.

A boundary that is considered to be already transgressed is the biochemical flow boundary, which specifically focuses around phosphor and nitrogen flows. These flows relate to food production operations. To feed the ever-growing human population of a fixed-sized planet, agricultural systems increasingly rely on fertilizers rather than on nutrients that are naturally available in the soil of ecosystems. These fertilizers contain phosphor and nitrogen, which are important raw materials that grass, crops, and other plants need to grow. The phosphor flow entering erodible soil and the nitrogen flow entering global cropland and grassland ecosystems have transgressed their boundaries, perhaps by more than 100 percent [543]. Transgression in relevant local ecosystems can be even more severe.

Nitrogen is needed for photosynthesis and is an important building block of proteins. Figure 10.3 provides a Shankey diagram showing the global nitrogen flows and how they have developed over the past 50 years as the global population grew from, roughly, 3 to 7 billion. It shows how the nitrogen volumes fed into global food production operations have more than quadrupled, and the majority of this growth is in the form of synthetic, industrially produced fertilizers (of which the use has grown by a factor of six) [330]. Moreover, Figure 10.3 shows that the majority of the nitrogen that enters the system is for crop production.

Of the 163 million tons used in 2009 for crop production, only 75 million tons were used for food for livestock and humans. Hence, the majority can be considered to have been wasted and remains in the cropland ecosystems, from where wind and freshwater may transport it to other ecosystems. These considerable yearly wastes negatively impact sustainability, for example, threatening biodiversity and freshwater availability, to mention two other safe operating space boundaries [322].

Over the past 50 years, population growth has not been the only driver of nitrogen use growth. The nitrogen growth rate has been double the population growth rate. Changes in diet toward increased consumption of animal proteins (especially

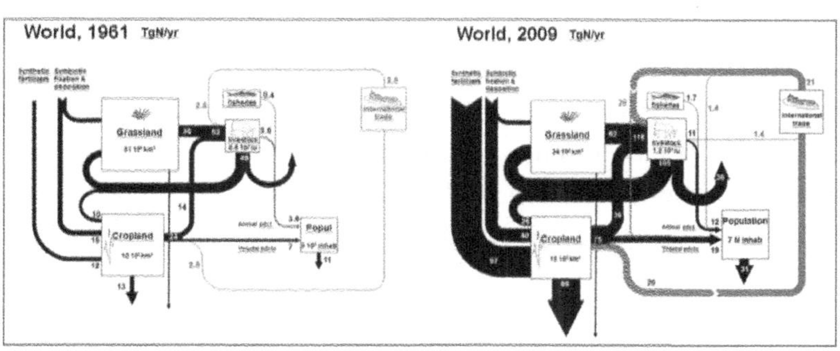

Figure 10.3. Nitrogen flows in the global food system 1961 versus 2009 [330].

ruminants such as cattle and pigs) have roughly caused the remaining nitrogen use to increase [330, 537].

The food supply chain, which transforms vegetable protein into animal protein, reaches humans at the top of the food chain less efficiently, as the more than four-fold increase in the raw material nitrogen only yields a tripling of human nitrogen intake at the top of the food chain. In 2009, 75 percent of the nitrogen contained in crops produced served to feed livestock, i.e., as input for the next stage of meat production operations in the global food supply chains [330]. As shown in Figure 10.3, the wasteful nitrogen use in the operations of the "livestock" stage also results in a substantial flow of nitrogen losses flowing into the ecosystems of planet Earth.

Leaner and more circular nitrogen supply chains are needed to bring today's global food supply chains within a safe operating space, and more so as further global population growth is expected. Chapter 11 discusses changes in production and consumption in the way we work and live that can lead to an effective transition toward sustainable nitrogen flows.

Phosphor is an equally essential raw material for plant growth, and the low availability of phosphor in soil can significantly limit plant growth [521]. Phosphor naturally occurs in soil and is replenished by weathering. In addition to naturally available phosphor, global food production operations rely on mined phosphor, which likely formed more than half of the approximately 40 million tons of phosphor entering the global soil in 2008 [94]. It has been estimated that one-quarter of phosphor is not consumed by plants but remains in the soil.

Soil phosphor leads to increases in phosphor transported to aquatic ecosystems in lakes, seas, and oceans, and these flows have tripled since the onset of the industrial revolution [94]. Through a process called eutrophication, phosphor increases in aquatic ecosystems have already resulted in dead zones, loss of fish species, and other forms of degradation of freshwater and coastal ecosystems [94, 322].

Let us now turn to one of the boundaries at risk of being transgressed soon: the CO_2 boundary that is linked so tightly with climate change. The boundary for atmospheric CO_2 is set to range between a minimum of 350 and a maximum of 450 ppm, well above the pre-industrial level of 280 ppm. Figure 2.3 from Chapter 2 shows how net emission rates have grown by a factor of more than thousands since the onset of the industrial revolution. This has caused atmospheric CO_2 levels to surpass the 400 ppm level around 2015 and to surpass the 420 ppm level in 2023 [146, 321, 322]. At current rates, we may therefore expect to transgress the 450 ppm boundary between 2030 and 2035 [522]. The boundary upper bound of 450 ppm corresponds to rises in average global temperature above which irreversible impacts on the ecosystems of planet Earth are highly likely to occur [322, 456].

The agricultural sector discussed in relation to the biochemical flow boundaries above, together with forestry and other land use, accounts for more than one fifth of 2018 net carbon emission rates, as illustrated in 1.1. Industry, when including the energy sectors, accounted for more than 45 percent of these emissions in 2018. The remainder is roughly divided between end consumers (the way we live) and the service industry, including transportation.

The impact of the increases in atmospheric GHGs goes well beyond temperature rises and impacts many other dimensions of sustainability. Climate change has altered marine, terrestrial, and freshwater ecosystems all around the world. It has negatively impacted biointegrity by causing local species losses, mass mortality events of plants and animals, and possibly the first climate-driven extinctions [456]. Climate change also has negative and sometimes irreversible impacts on planetary ecosystems through more frequent and more severe extreme weather events, including droughts, wildfires, heatwaves, and cyclones [456]. These impacts can be particularly severe in local climate-sensitive ecosystems, such as mountain top ecosystems and polar ecosystems, and in the species living in these ecosystems.

Climate change is also impacting grassland and cropland soil yield, thus impacting the effectiveness of agricultural operations. This may in turn impact the use of more limited available freshwater and biochemicals, thus contributing to further transgressing the local and global biochemical flow boundaries of a safe operating space [322].

The aforementioned impacts of climate change are not intended to be exhaustive but rather highlight some of the most important observed consequences. Scenarios that include potential future impacts as well as mitigation and adaptation efforts are discussed in Chapter 11.

10.2.2 Social Inclusion

We now reflect on present-day sustainability regarding social inclusion on the basis of the dimensions introduced in Chapter 2. We first reflect on the findings from previous chapters and related recent updates and then connect with the impacts of the (possible) transgressions of ecological boundaries discussed above.

Before advancing, let us consider the hypothesis that operations are only indirectly related to social inclusion, i.e., to intragenerational and intergenerational equity in meeting the needs of present and future human generations. To support this hypothesis, one might argue that operations simply regard the value creation processes and the value produced, e.g., in terms of GDP, and are not concerned with a socially inclusive allocation of the value created among present and future generations of humankind. After all, the global per capita GDP produced by human operations is far above all commonly considered poverty limits as expressed in per capita income. Moreover, decisions about the spending of income by individuals, households, and populations relate to consumption, not production.

These arguments are overly simplistic. Previous chapters have shown that operations can compromise the ability of present and future generations to meet their needs, especially when ecosystems change. The productivity increases of hunter-gatherers in the hunting of large animals were such that these animals became extinct, causing their operating model to become unsuitable for future generations of Homo sapiens and other Homo subspecies, some of which went extinct. These effects of human operations likely contributed to the onset of the agricultural revolution (see Chapter 5). The operating models of the agricultural revolution, in turn, relied heavily on slavery and other forms of coerced labor and have negatively impacted contemporary populations of hunter-gatherers.

British and Spanish wood harvesting practices to build ships for their overseas operations and to provide energy for their homes and factories caused deforestation. These unsustainable operations caused changes in the operating model, such as moving ship building practices overseas—where they continued to cause unsustainable deforestation—and the use of alternative energy resources such as coal, as described in detail in Chapter 7. The same chapter also illustrates that much of the GDP growth in the next industrial stage was realized by operating models that exploited indigenous populations, imported slaves, and *raged a war on the working population*" in industrialized homelands. The operating models were founded on the systematic and extensive use of child labor, slavery, and other forms of coerced labor, practices that the SDG agenda still seeks to eliminate.

Humankind has a long track record of deliberately designing and practicing operations that are presently viewed as lacking social inclusion and unsustainable. Moreover, humankind is continuing several of these unsustainable practices on large scales. At present, 160 million children are working and providing cheap labor instead of being in school, thus compromising their ability to meet their future needs [500]. More generally, many workers earn incomes that are insufficient to rise above poverty and maintain their personal health and the health and education of their family members, whether in agriculture, industry, or the service sector.

The competition between global value chains still causes operations to be relocated to places where salaries and other costs, such as environmental protection, are the lowest and often unsustainably low [308, 309]. For instance, some evidence points to the relocation of operations in global supply chains to avoid costs of energy and carbon emissions, thus increasing the carbon footprints of products and services produced despite regulations implemented to prevent such "*carbon leakage*" [245, 463, 632]. At the other end of some global value chains, the operations of recycling electronic waste are relocated to African countries that have less restrictive regulations and practices, negatively impacting local human health and ecosystems [387]. The low cost of these operating models often forms an essential element of their competitive advantage. The operating models of these businesses may be financially unsustainable when accounting for sustainable resource cost levels, including wages that generate a socially inclusive income.

The design, control, and improvement of operations by humankind are thus core to extanting problems to meet the needs of present and future generations in a socially inclusive manner. This applies both to the operations of work and to the operations of life, as highlighted by SDG 12 on Sustainable Consumption and Production. If consumers regarded goods and services produced by unsustainable operations as lowly valued and disregarded purchasing them as inputs for their operations of life, these goods and services would no longer be produced.

Zooming out for a global perspective on social inclusion in relation to operations, let us first recall from Chapter 9 that present income inequalities are close to the levels of the early days of the industrial revolution despite the exponential growth of global (per capita) GDP and GNI over the same time period [103]. Income inequalities between countries have decreased, while inequalities within countries tended to increase [103]. Thus, improvements in operations that drove global and national GNI increases have not translated into income inequality reductions from a global perspective, and the opposite has typically happened at the country level.

The latest evidence reveals that poverty and extreme poverty persist in the least developed countries and elsewhere and that humankind remains far from the social inclusion objective of leaving no one behind [323]. Across the globe, hundreds of millions of people live on incomes below the extreme poverty threshold of a purchasing power equivalent of 1.90 1990USD per day. In absolute terms, their number has been relatively stable over recent decades. Against a quickly growing global population, this implies the percentage of the global population with incomes below this limit has decreased rapidly and in fact dropped below 10 percent for the first time around 2015 (see also Figure 2.2). The rate of decrease has slowed down since then, and the COVID-19 pandemic has likely resulted in an increase of almost 100 million people who have fallen into extreme poverty since the onset of the pandemic [323].

Taking a less extreme limit to define and measure global poverty, a per capita income value of approximately 7 2017USD is the lowest value that yields the absolute population with incomes at or below this line at the onset of the pandemic, i.e., in 2019, to equal the 1990 level of 3.7 billion. This number initially increased with global population growth after 1990, reaching a maximum level of 4.1 billion, but has decreased toward a more equitable distribution since [36]. In relative terms, the population living on less than 7 2017USD has decreased by more than 30 percent since 1990.

Altogether, it appears that the increases in global per capita GDP and GNI leave a relatively constant-sized population of the poor behind, while a growing population enjoys income increases. This development is also reflected in a global per capita GNI increase roughly from 27 to 46 international 2017USD per day in the past three decades.

The quantitative analysis above should, however, be considered with caution as much of the value produced by operations—and received as income—is not reflected in formal GDP and GNI statistics, which fail to record (much of) the informal, "shadow," economy [37, 571, 604]. Let us recall that around 60 percent of the global workforce is not (fully) included in the formal economy but rather informally employed and thus contributes to operations that are unlikely to be reflected accurately in GDP and GNI, if at all [67, 571].

Informal operations are often conducted to mitigate the consequences of not being (fully) included in the formal economy. The informal operations may serve to avoid poverty and subsequent struggles to meet sustainability boundaries in other social inclusion dimensions affected by poverty, such as access to food, sanitation, health services, education, housing, and energy. Informal operations are especially common in the service sector, which has outgrown other sectors in recent times, as covered in Chapter 9. The relative importance of income from the informal economy tends to be greater for women, for the population of less developed countries such as Sub-Saharan Africa and South Asia, and for the poor [578].

The social inclusion dimension of well-being may be considered to be as least as relevant as income. The two dimensions are highly correlated at low income levels but decreasingly so as income increases [571]. Focusing on inclusive, equitable, and well-being therefore prioritizes income equity even more than focusing on income as a measure of social inclusion itself.

Let us now turn to considering how the recent ecological sustainability challenges that result from human operations affect social inclusion. The doughnut model was developed to address these relationships [472, 473]. It connects an outer ring of boundaries defining an ecologically safe operating space for the planet with an inner ring of social inclusion boundaries derived from internationally agreed minimum standards for human well-being, as established in the SDGs [473].

Before looking into the impacts of present and future boundary transgressions, it is appropriate to establish that they are unevenly caused by different human societies and different social groups [543]. Correspondingly, the wealth benefits that these transgressions have brought are also unevenly distributed, both socially and geographically. The current wealth of nations and populations is, for instance, associated with GHG emissions since the onset of the industrial revolution. Such uneven distribution of the causation and benefits of unsustainable operations is expected to continue, further underlining the essential role of operations in social inclusion shortfalls [543].

A main pathway from the outer ring of planetary boundaries to the inner social inclusion ring is via reductions in the yield of operations. Climate change in the forms of heat and drought, biochemical flows that cause infertility of land, floods, wildfires, et cetera, reduces operating yields and subsequently income [456]. In fact, this pathway and closely related ones lead into a much broader set of social inclusion domains, such as poverty, limited access to work, food, water, housing, and energy [456].

The incomes of populations with direct return operating models, such as pastoral people and small-scale farmers, are most directly impacted by impacts on operating yields and therefore especially vulnerable to the transgression of a safe ecological operating space. For instance, well-documented evidence describes the negative impacts of climate change on the livelihoods of pastoral people in Algeria, rural farmers in the Ningxia autonomous region in China, and female farmers in Tanzania [331]. The latter case study illustrates the overrepresentation of women among the global poor and, hence, the gender inequality in the effects of transgressing ecological boundaries. More generally, the impacts on income and food security are greater for populations at low and mid latitudes, low-income households, indigenous peoples, minority groups, small-scale producers, and fishing communities [456].

Likewise, women, children, the elderly, indigenous people, low-income households, and socially marginalized groups are most vulnerable to the negative health impacts caused by climate change [456]. More specifically, the negative impact on livelihoods has been associated with mental health problems, reduced well-being, and violence [456]. Moreover, difficulties in access to food have caused malnutrition, heat and drought have caused mortality, changes in climate have increased the prevalence of infectious diseases, and air quality has increased the prevalence of and morbidity from respiratory diseases. This provides further evidence that the wealthiest populations whose operations have contributed most to causing negative impacts on sustainability are not among the populations most vulnerable to these impacts. By contrast, the disproportionately affected present-day hunter-gatherers, pastoral people, and small-scale farmers have sustainably practiced direct reward operating models for thousands of years and well within the boundaries of a safe planetary operating space.

10.2.3 Economic Development

Chapters 1 and 2 discussed how economic growth is one of the three domains of sustainable development [398]. Economic growth for the poor is explicitly elaborated in SDG 8, Decent Work and Economic Growth, and considered essential for the SDGs in the social inclusion domain, such as SDGs 1–4, No Poverty, Zero Hunger, Good Health and Well-Being, and Quality Education [500].

GDP is the classic measure of economic growth. Moreover, GDP is a measure of the total value created by operations and is therefore also most fitting for our research aims. Hence the important role it has played in the preceding chapters.

Chapter 2 also reflected on the shortcomings of GDP as a measure of economic development, which are especially relevant from a sustainability perspective. The exponential economic growth that the preceding chapters describe in detail results from human operations that cause unsustainable transgressions of the boundaries of a safe and just operating space [163, 487, 518, 553]. Present GDP growth compromises future GDP growth and the needs of future generations.

In 1996, the United Nations Development Program (UNDP) identified five types of harmful growth, among which are ruthless growth (that increases inequality) and futureless growth (that depletes national resources) [125, 566]. Growth that damages ecosystems and their future value creation potential is a form of futureless growth. This observation has led to "greened," or "environmentally adjusted," GDP measures, which account for changes in natural resource reserves, such as fossil fuels, metals, and minerals, and land resources in forestry or agriculture on the one hand, and for the impacts of carbon emissions and other forms of pollution on the other [125, 316, 571]. These adjustments are closely related to the inputs and outputs of operations and better incorporate their impact on the needs of future generations. A related alternative approach is to consider the resource intensity of GDP, i.e., the resource use divided by GDP [291].

Environmentally adjusted measures still leave the social inclusion dimensions of a just operating space unaddressed. Because of this unsustainable characteristic and lack of coherence with other sustainability measures, GDP growth is neither a suitable nor an accurate measure of sustainable development. We now therefore continue the exploration of alternative economic development measures, as already initiated in Chapter 2.

A first alternative has already been adopted by the UN. SDG 8 limits the consideration of GDP as a sustainability measure to GDP growth in the least developed countries. This more specific focus reflects the need to achieve economic progress for the poor. (The average GDP growth of the least developed countries since formulating the SDGs has been consistently below the 7 percent target of the SDG agenda [500].) The SDG agenda also includes targets on two related measures for the operations of work in the least developed countries regarding labor productivity and employment rates in the least developed countries. (The labor productivity in these countries has been struggling to increase since the onset of the COVID-19 pandemic [500].) Unfortunately, however, we have already provided recent evidence above of GDP increases realized while diminishing equity. Thus, while GDP growth for low- and middle-income countries may be necessary for sustainable development, it is not an indicator of sustainable development.

A second alternative is to replace the economic measure (per capita) GDP with the measure (per capita) gross national income (GNI) from the social inclusion domain. This repairs the shortcoming that the value of goods and services produced may not end up as income for the individuals or populations producing it. It reflects that "*human development is the end—economic growth is a means*" [566]. Income is indicative of sustainability if it develops with intragenerational and intergenerational equity. Income and income distributions can replace GDP as a measure of the value created by operations that is consistent with the social inclusion indicators of the doughnut model for a safe and just operating space. In the remainder, we therefore disregard GDP (growth), effectively giving up on target 17.19 of SDG 17, which aspires to "*build on existing initiatives to develop measurements of progress on sustainable development that complement gross domestic product*" [518].

For completeness, let us briefly discuss more advanced "beyond GDP" measures as recently developed [125, 291, 316, 566, 571].

The Index of Sustainable Economic Welfare (ISEW) is a seminal measure beyond GDP and is based on income (GNI) rather than production (GDP) [118]. The Genuine Progress Indicator (GPI) is based on the ISEW and includes components typically not considered in GDP, such as the value created by the informal economy and the costs of environmental deterioration, crime, and pollution. Moreover, it takes income distribution into account [118, 316]. It is intended to be a measure of current welfare rather than of sustainability [118, 316]. The same applies to GPI 2.0, which nevertheless is the only measure to include a component for welfare losses for future generations [548].

Interestingly, per capita GPI has been shown to be highly correlated with GDP, up to a value of around 7,000 2005USD, which is only slightly below the global per capita GDP in 2005. It has also been shown to be negatively correlated with GDP for higher GDP values [316]. The same authors show that global per capita GPI has been relatively stable over the period 1970–2005, while global per capita GDP almost doubled. These findings further invalidate GDP as a sustainability indicator.

Another set of alternative measures considers constructs such as welfare, well-being, quality of life, and happiness [125, 571] rather than income. These measures are particularly relevant for populations with higher per capita GDPs, for whom the correlation between growth in terms of these measures and per capita GDP growth is weak. When reported, these measures typically represent development relative to the past year rather than well-being effects for future generations.

The HDI has been proposed and reported yearly by the UNDP and is defined as the average of three measures for health, education, and economic attainment

[566]. Initially, economic attainment was measured in per capita GDP, health was regarded as life expectancy at birth, and educational attainment was measured in terms of literacy. In a later stage, economic development became measured in per capita GNI and subsequently inequality adjusted per capita GNI [17, 566]. Further adjustments, which also incorporate health inequalities and educational inequalities, have resulted in the inequality-adjusted HDI (IHDI), as also adopted by the UNDP [121, 263]. Moreover, with a view toward environmental protection and sustainability, a planetary pressures-adjusted HDI has been developed and reported, which adjusts for material use (footprints) and GHG emissions [120, 121]. The definitions of all HDI variants and their components are cross-sectional, i.e., relative to other countries in the same year, and hence for the present population. HDI and variants therefore give little insight into longitudinal development, particularly into the development of future generations.

10.3 The 4th Industrial Revolution?

Let us now start turning to the question of how the 4th industrial revolution can help resolve today's sustainability challenges. To achieve this purpose, it is first necessary to define and understand this revolution. What are the fundamental alterations it entails? What are the disruptive changes it brings that might cause the future to be "*unlike anything humankind has experienced before*" [517]?

10.3.1 Introducing the 4th Industrial Revolution

Chapter 1 presented the viewpoint "*that we stand on the brink of a revolution that will fundamentally alter the way we live, work, and relate to one another*" [517]. This citation from the seminal source on the 4th industrial revolution built on the German *Industry 4.0* initiative from earlier in the same decade, which aims at a planned 4th industrial revolution [329]. On purpose, the naming is "*reminiscence of software versioning*" and suggests an evolutionary progression [329]. Industry 4.0 refers to the production of goods and can therefore be positioned in the secondary sector. The underlying fundamental concepts are as follows:

1. Smart Factory,
2. Cyber-Physical Systems,
3. Self Organization,
4. New Systems in Distribution and Procurement,
5. New Systems in the Development of Products and Services,
6. Individualized Product and Service Development,

7. Adaptation to Human Needs, and
8. Corporate Social Responsibility.

The term Cyber-Physical Systems refers to the merging of physical and digital representations of products, resources, and systems [329]. The state of these objects is then defined by a combination of physical and digital parameters. The digital representation may also be connected to the Internet, thus joining the Internet of Things (IoT).

Industry 4.0 has also become the name of a corresponding project implemented by the German government [329]. Various other national and international projects akin to Industry 4.0 have been initiated, some of which involve the primary and tertiary sectors. Chapter 5 already introduced a present-day agricultural revolution named Agriculture 4.0 [490]. Likewise, it has been posited that the same and similar technological advancements that drive Industry 4.0 "*will transform virtually all service sectors*" and "*lead to rapid innovation that can dramatically improve the customer experience, service quality and productivity, all at the same time*" [613].

The European Commission has already initiated Industry 5.0, which intends to complement and extend Industry 4.0, emphasizing "*aspects that will be deciding factors in placing industry in future European society*" [130, 627]. In 2016, the Japanese government adopted the Society 5.0 initiative, which refers to a "*society built upon Society 4.0, aiming for a prosperous human centered society,*" where Society 4.0 refers to the society resulting from the 3rd industrial revolution [209]. Initiatives numbered 6.0 are in the making.

The successors of Industry 4.0 tend to be more extensively and explicitly related to sustainability challenges, and in particular to the SDGs [518]. Several of these successors refer explicitly to human-centeredness and "society-centeredness," sometimes using different wording [130, 209, 627].

The appearance of numbered successors of the first industrial revolutions and of related numbered initiatives indicates that the fourth, fossil-fueled machining stage of Figure 10.1 is evolving and making subsequent threshold advancements steps that bring further fundamental alterations to the operations of work and life.

Such advancements may be difficult to identify and appreciate as threshold developments while they are evolving. Agriculture, the steam engine, and telegraphy are all examples of threshold developments that evolved over a long period of time until they started to have a revolutionary impact on human operations. Thus, it may well be the case that we are presently overlooking some of the evolutionary developments that will, in hindsight, turn out to have been the seeds of revolutionary advancements. On the other hand, some of the threshold developments have indeed resulted from consistent and purposeful effort, as has been the case for the automobile and the mainframe computer (see Chapters 8, 9). Hence, in careful modesty, one may realistically aspire to grasp the threshold developments brought forth by the 4th industrial revolution, Industry 4.0 and its successors [209, 307].

The 4IR has been posited to be "*unlike anything humankind has experienced before*" in terms of "*scale, scope, and complexity*" [517] as it blends physical, digital, and biological changes to the way we work and live [517]. A more explicit definition states that "*the Fourth Industrial Revolution can be described as the advent of cyber-physical systems involving entirely new capabilities for people and machines,*" which may involve biological changes [144].

Present-day 4IR developments in non-biological physical technology, digital technology, and combinations of both can be viewed as evolutionary advancements from previous industrial revolutions. They bring threshold developments, but likely without defining a new, fifth stage of the timeline of operations. These changes can significantly alter the way we work and live, as has been the case on many occasions since the onset of the industrial revolution, as elaborated in Chapter 9. In this chapter, we consider the effects on present operations. In the next chapter, we discuss future, possibly more revolutionary, developments, including the much-debated topic of the future of work.

Current 4IR advancements have already caused a widening of income inequalities between those working in routine jobs requiring less (computer) skills and those working in more complex jobs requiring high computer skills. However, it appears that in the decades before and after the turn of the millennium, less than 5 percent of the workforce was employed in the ICT sector itself or in jobs newly created by technological advancements (see again, [55] and the references therein). A much larger number of jobs in the primary, secondary, and even tertiary sector have on the other hand become obsolete because of the increased capabilities and reduced prices of new (information and communication) technology. This caused a relative increase in the size of the workforce working in low-skilled jobs in technologically stagnant sectors, for which operations were affected less by technological advancements. As explained in Chapter 9, technological stagnancy mostly occurs in service operations, such as healthcare [48]. From a global perspective, recent technological advancements appear not to have caused employment rate reductions.

The increased effectiveness of production factor technology, compared to human labor, has yielded investment in ICT more attractive. This caused the labor share of income to drop by approximately 5 percent around the turn of the millennium while the incomes of those investing in capital increased. The technological advancement thus likely contributed to a further widening of income inequalities [55].

It follows from the above that non-standardized operations that are not executed in large volumes are less likely to see human activity replaced by technology. This applies, for instance, to operations requiring creativity, social and emotional skills, and analysis and problem solving ("*perception and manipulation*") [208]. At the same time, it has been observed that as technology advances, operations that were previously considered non-standardized or low-volume can become standardized and prone to automation. This may hold especially true for advances in artificial intelligence (AI) and, more specifically, machine learning, which aim to equip technology with various forms of intelligence, among which are problem-solving skills, social skills and forms of creative intelligence.

One might hope that these early effects of the nascent 4IR also apply to the operations of life, helping humans to spend less time on repetitive tasks requiring low skills (such as dish washing [55]) outside of work and to spend more time on activities that call up on their creative, social, and emotional abilities. Others, however, might rather hope to live without such technological progress. Indeed, some human populations prefer to continue to live as hunter-gatherers, or by practicing small-scale farming. Reduced technology adoption may also imply reduced carbon emissions and help prevent transgressions of other planetary boundaries. Conversely, the advances of the 4IR can easily promote more energy-intense operations of work and life and increase pressure on the planetary boundaries.

From the above, it appears that the 4IR has mostly had a negative impact on environmental protection and social inclusion until present. Even when promoting productivity, it has tended to reduce income from labor, especially for those with lower incomes. Furthermore, the technology of the 4IR may contribute to global increases in fossil fuel use and shift income toward the part of the human population with a larger carbon footprint.

These unsustainable impacts are not intrinsically associated with the nature of the technologies in the 4IR. They rather follow from how these technologies are being deployed and how operating models and business models are adapted to take advantage. This viewpoint is further elaborated in a recent review of the potential contributions of AI to the SDGs [588]. It finds that AI can indeed act as an enabler for the vast majority (71–79 percent) of the SDGs [588]. In fact, AI can already be deployed to identify evidence-based interventions toward achieving the SDGs [454].

AI can also hinder progress toward the SDGs, as may be the case for 21 to 35 percent of the SDGs [588]. Because of its reliance on large data sets and computational capacity, AI-based operations are energy-intensive and can thus easily contribute to GHG emissions. The resource and energy intensity can also cause access to be limited in less developed areas or for certain sociodemographic populations. This complements the findings above that less skilled people are more likely to see their incomes negatively affected because of such technological advances. AI, in particular, can also "learn" discriminatory and other forms of unethical operational practices and subsequently promote inequality [588].

10.4 Operations Management Reflections

Two straightforward reflections regarding today's operations management are that (1) it has played a pivotal role in enabling humankind to live with more than 7 billion people on planet Earth and so far with ever-increasing average incomes, while along the way (2) humankind has managed their operations toward and across some of the boundaries of a safe and just operating space for planet Earth and society. Operations management thus also plays a pivotal role in possible harmful effects of operations on planetary ecosystems, some of which are irreversible at timescales meaningful for humankind and/or are co-causing a variety of social inclusion shortfalls, as further elaborated below.

Operations management has played a pivotal role in realizing economic growth as humankind (re)designed, controlled, and forever improved its operations while adopting the technological innovations documented in the preceding chapters, particularly the innovations of the four industrial revolutions. Operations management translated entrepreneurship and management of business growth into processes creating value in the form of products and services. The fossil fuels used to operate the technologies of these enterprises presently account for around 80 percent of global GHG emissions (1.1).

Business does not form a class of tangible ecosystem elements of its own. Businesses are social structures created and applied by humans to organize their work. These businesses form the value chains supplying the products and services for the operations of life. The carbon footprints of businesses are ultimately included in the carbon footprints of consumers. More generally, the ecological and social footprints of business operations are ultimately included in the corresponding consumer footprints. This broader view positions business operations management as a function of SDG 12, sustainable production and consumption.

SDG 12 brings us to the perspective that operations management forms a lever to adjust operations or redesign new operations in response to demand for (more) sustainable products and services. This demand can be made by direct customers and by customers at the end of the value chain, possibly recycling the products to form a value circle.

Changes in legislation and regulation can also force the adoption of more sustainable operations management practices. In both cases, these requirements can travel up and down value chains, where, for example, regulations in one country require a supplier in another country to manage changes toward more sustainable operations. Examples of operations management practices in response to regulatory requirements and in interaction with value chain partners are documented in various recent reviews [25, 576, 591].

Some evidence suggests that such extrinsic motivators are not among the most effective mechanisms to improve the sustainability of operations [217]. Management efforts that leverage intrinsic motivation are increasingly common. For instance, current initiatives for sustainable healthcare and green hospitals are typically initiated by management and employees and not necessarily reflective of customer priorities or governance regulations [616]. Intrinsically motivated, supply-driven initiatives can also spread within industries or across industries, as is, for instance, the case with certification initiatives and related forms of value chain sustainability governance [69, 417].

Progress toward resolving the present sustainability challenges is likely to be most substantial when the intrinsic and extrinsic motivators are aligned. Below, we more deeply reflect on current operations management challenges and how to align solutions for each of the three domains of sustainability.

10.4.1 Perspectives on Current Operations Management for Economic Development

When setting key performance objectives, operations management practices still seldom look beyond economic objectives such as the minimization of cost or the maximization of revenue or profit [576]. These practices reflect that operations are primarily viewed as cost centers, revenue centers, or profit centers, depending on the business and the financial business logic adopted.

Cost centers, for instance, may be managed with a primary focus on efficiency and disregard sustainability-related performance dimensions beyond ensuring compliance with corresponding laws, regulations, and a selection of other standards. We have seen in previous chapters that even adherence to laws, regulations, and standards may be sacrificed in pursuit of financial performance [355, 532]. The efficiency focus has been of prime importance since the industrial revolution and is for instance highlighted in Chapter 8 when considering Frederick Taylor's operations management practices to quadruple the load of pig iron workers carried per day to 47.5 tons, and in Chapter 9 when telephone switchboard operators were expected to connect a call every 3.5 seconds. It is also reflected in the never-ending pursuit of waste elimination by Ford and Toyota, illustrated in the same chapter, and is still at the core of lean management as widely practiced today.

The more recent view of operations as a set of value-creation activities, or a *value stream*, has led to viewing operations departments as revenue centers or profit centers. This view adopts a perspective of operations for *sustainable competitive advantage* and profitability, in which sustainability refers to a strategic, long-term business horizon rather than to effects on society or the planet [455]. This view steers operations management away from a cost focus and the corresponding "*race to the bottom*" of competing on cost and recognizes operations as the activities by which distinctive value can be created [455]. Still, a focus on competitive advantage to create value for customers and shareholders may leave social and environmental sustainability-related performance objectives to be compliant with laws, regulations, and other standards rather than being actively pursued [576].

In recent times, profit margins have increasingly benefited from the *economies of scope* attainable by technologically advanced, automated, high-volume production of a variety of products and services. The operations management practice of *mass customization* combines the low cost of automated high-volume operations with the value creation of customization. In recent decades, the adoption of mass customization in global value chains has enabled operating in the upper right-hand corner of the product process matrix depicted in Figure 5.3, moving orthogonally upward from the diagonal that has defined the feasible operations management space in preceding chapters.

This trend toward mass customization has importantly benefited from the principle of *postponement*, in which the customized, and therefore less standardized and typically more costly, operations are postponed as far downstream in the value chain. In many contemporary value chains, these downstream high-value-added operations are conducted close to well-resourced consumer markets in developing countries. These downstream operations may be less standardized and provide jobs for skilled workers. Standardized, low-skilled upstream operations are often offshored to locations where costs of labor and other resources are low [81].

An important characteristic of the emerging (artificially) intelligent technology of the 4th industrial revolution is its ability to substitute human operators for less standardized tasks and master the complexity of high-volume operations delivering customized products and services. Robots work on assembly lines producing high-segment customized cars, pick millions of customer orders in the warehouses of web shops for next day-delivery, and "man" the customer contact centers of multinational service organizations. These Industry 4.0 advancements may reduce investment in upstream manufacturing in developing countries with low human resource costs, as further elaborated in the next subsection [182].

The relocation of manufacturing plants that leverage Industry 4.0 technologies replaces human labor with technology. Moreover, technological advancements such as 3D printing, drones, and other forms of unmanned automated vehicles can facilitate the automation of downstream, possibly postponed, customized tasks that have traditionally required more highly skilled human operations [375, 477]. This automation can cause the deskilling or disappearance of jobs. The COVID-19 pandemic may, for instance, have accelerated drone-based delivery in the last mile of pharmaceutical value chains, an area for which evidence of successful implementation had previously remained modest [248, 370, 375].

Figure 10.4. Health service robot at work in Japanese elderly care [485].

The developments toward the implementation of 4IR technologies downstream in the value chain extend to the service industry. As observed in Chapter 9, the 4th industrial revolution also brought a service revolution [613]. Early evidence indicates service (ro)bots are effectively starting to replace human service operators in tasks requiring advanced cognitive skills and modest social and emotional skills and complement human operators in service delivery tasks requiring advanced cognitive and emotional skills [353, 613]. Provisioning of health services for the growing population of the elderly has received special attention while attempting to resolve the technological stagnancy of care tasks (see Figure 10.4).

With notable exceptions such as in rural healthcare and in elderly care, the operations management efforts to adopt novel technologies have mostly targeted economic development in the classical sense of GDP growth. The resulting cost reduction and efficiency improvement efforts may subsequently free up resources to be deployed for value creation elsewhere. For instance, mass customization enables efficiency improvement upstream along with increased revenue from more valuable products and services downstream. It thus frees up resources upstream and generates financial resources downstream to subsequently invest in the next round of technology for further cost and profit improvements. The aforementioned decrease in the contribution of labor to global income and the corresponding increase in income from investment in technology align with this development.

It becomes clear that the economics and operations management practices of business as usual may be detrimental to social inclusion and environmental protection. For sustainable development, it may be necessary to redefine the values and objectives of operations management. When aligning with the SDGs and with the purpose to contribute to staying within the boundaries of a safe and just operating space, operations management objectives broaden, from a focus on profit and shareholder value and a focus on revenues and customer value, to including values of employee income and well-being, providing decent work (SDG 8), and valuing the needs of present and future generations of humans and other species in the ecosystems affected by the managed operations. Operations management as usual may work against sustainability objectives as it tends to decrease the income of less skilled workers, who are already in lower-income jobs, and widen income inequalities and other social inequalities while increasing demand for nonrenewable (energy) resources.

Our analysis then duly shifts toward alternative operations management objectives in the next subsection, which regards socially inclusive operations management practices.

10.4.2 Perspectives on Current Operations Management for Social Inclusion

Operations management practices can be socially inclusive for those directly involved in the operations of the value chain and more broadly for all stakeholders affected by these operations, among which are customers, value chain partners, employees, neighboring communities, et cetera, including those who are indirectly affected by the environmental impacts of the operations.

SDG 8 connects economic growth to the operations of work in a socially inclusive manner by aspiring to promote sustained, inclusive, and sustainable economic growth, full and productive employment, and decent work for all [169]. According to the International Labor Organization, decent work involves, among other aspects, "*opportunities for work that is productive and delivers a fair income, security in the workplace, . . ., and equality of opportunity and treatment for all women and men*" [420]. This relates SDG 8 directly to the social inclusion dimensions of income, work, health, and well-being, and via income to dimensions such as poverty and access to food, among others [473]. For operations management, it brings the challenge of designing the operations of the value chain so as to equitably provide fulfilling jobs and decent incomes. At present, this brings the additional challenge of leveraging the 4IR in service of such decent work and stopping it from being a counterforce that renders jobs obsolete and increases income inequality.

These aspirations are not new, and Chapter 8 already outlined how Frederick Taylor, Lilian, Frank Gilbreth, and Henry Ford sought to create jobs that avoided hard work and fatigue and maximized prosperity for employees and employers alike during the 2nd industrial revolution. We may also recall from the same chapter that they advocated efficiency gains to translate to salary increases of up to 100 percent, as realized by Ford in 1914 [202, 222, 551]. Taylor explicitly advocated collaboration between management and workers, recognizing the pitfalls of operations management advances coming "*from the side of management.*" Lilian Gilbreth explicitly introduced operations management practices that promote welfare and well-being, including the mental and moral well-being of employees [223].

During the 20th century, the ideas of these early operations management experts spread across many of the countries that developed and industrialized relatively early. In recent decades, however, the globalisation has often taken the form of relocating, offshoring, and manufacturing operations to developing countries with lower wages. While potentially socially inclusive, this offshoring has also relocated labor to contexts in which work and job design have not yet adopted the social inclusion standards of developed countries. Thus, the theme of socially inclusive operations gained renewed interest.

Evidence suggests that multinationals typically provide higher incomes and better working conditions than domestic firms in the same locations [81]. At the same time, evidence suggests this is not the case for domestic suppliers of these multinationals, whose operations are outsourced [355]. The competitive demands for efficiency and return on investment in global value chains can easily push them toward evasion of standards, even in audited and certified global value chains, which unfortunately has led to a variety of adverse outcomes for workers, including suicides and events with large numbers of casualties [305, 355, 532].

Chapters 5 and 7 have provided historical examples of the lack of social inclusion in the primary sector, especially in relation to the harvesting of cotton, i.e., at the source of global clothing value chains. Global agricultural value chains continue to be of greatest importance in today's society for the provisioning of food products such as soya, cacao, coffee, sugar, and vegetable oil. The lack of social inclusion in agricultural operations at the source of these global supply chains remains a concern in current operations management. While documented examples of socially inclusive global supply chain operations management practices that benefit small-scale farmers and workers at larger farms exist, the evidence suggests that the forces exerted by global markets and global value chain management practices are typically at odds with the social inclusion of small-scale farmers and agricultural workers [220, 442, 585]. With the exception of some examples evidencing that the adoption of mobile phone-based technological advancements can empower small-scale farmers and improve their income, current evidence of the benefits these technologies bring for small-scale farmers is scarce and inconclusive [49, 187, 369]. For the global population of 2 billion living from small-scale farming, one may hope that further technological advances and cost decreases will help to overcome the barriers to effective technology deployment in developing countries discussed in Chapter 9 [187].

Because of trade wars, political conflicts, the COVID-19 pandemic, natural disasters, and transportation difficulties, offshoring of manufacturing has been reconsidered, and the reverse development of reshoring (or backshoring) has recently gained momentum [441]. The extensive labor operating models of the 4th industrial revolution can further stimulate reshoring as they reduce the benefits of low labor costs [41, 219]. Lights-out manufacturing, i.e., manufacturing in facilities without regular lighting as there are no human operations, serves as the extreme zero-labor example [182]. While this may modestly increase opportunities for decent work in the (developed) countries to which operations are reshored, such manufacturing reshoring is expected to eliminate work and income for the less skilled in developing countries, with women disproportionately affected, and hence to be detrimental to reaching the SDG 8 objectives [109, 219]. Thus, a main challenge for current operations management is to leverage 4IR technologies to promote socially inclusive work in the global value chains, with a corresponding focus on the lowest-skilled and lowest-income populations.

Let us close this subsection by providing some examples of operations management examples showcasing successful socially inclusive operating models. In Brasil, Botswana, and China, the social enterprise Solar Ear produces very low-cost hearing aids with batteries that can be recharged by a dedicated solar-powered charging station [559]. It employs deaf workers, who are otherwise disadvantaged at the labor market, and enables a population of poor people with hearing loss to overcome their disability [421]. In so doing, it promotes social inclusion for employees, customers, and their communities, while also promoting environmental sustainability through its solar-powered product design.

The Aravind Eye Hospital highlighted in Chapter 9 also provides access to decent work for local workers while providing socially inclusive access to cataract surgery [273]. The low-cost operating models, integrated with a low-cost product design (viz., the lenses), are essential elements of their socially inclusiveness. The services of Aravind are presently being complemented by the artificially intelligent mobile phone app e-Paarvai, which enables volunteers to screen and diagnose cataract in remote areas lacking access to ophthalmologists [580]. E-Paarvai creates value as it improves health, well-being, and income for rural patients suffering from cataract, upskills volunteers to diagnose the disease, and promotes the effectiveness of ophthalmologists, while reducing the costs.

10.4.3 Perspectives on Current Operations Management for Environmental Protection

For some of the planetary boundaries, the relationship between operations and operations management on the one hand and environmental protection on the other are straightforward. For instance, forestation and deforestation are operations that directly relate to land system change. The operations of value chains that may involve land system change can then be managed with the objective of minimizing unsustainable surface area changes, such as the hectares changed from tropical rain forest to grassland. Conversely, sustainable operations management can seek to maximize land system changes toward more sustainable types, such as hectares changed from cropland to boreal forest.

Forest ecosystems play an essential role in staying within the CO_2 boundary, as forests relatively effectively remove CO_2 from the atmosphere. Operations in value chains that may cause land system changes thus also may induce changes in GHG emissions and removals, as further considered below when discussing the CO_2 boundary.

The environmental sustainability dimensions of land system change and fresh water also relate to other boundary dimensions, and in ways that may be harder to relate to the operations managed within an organization or value chain [322]. Chapters 3 and 4 describe how land systems change from forests to croplands and grasslands, which reduces the habitat of beavers and threatens their population size and genetic diversity. These effects of human ecosystem engineering on beaver populations subsequently impact other species, as beavers are effective ecosystem engineers themselves. Obviously, such follow-up effects on biodiversity can be difficult to attribute to specific operations or operations management decisions.

Human ecosystem engineering also brings habitat changes for bees, which are one of the drivers of recent bee population decline and bee subspecies extinctions [237, 617]. Biochemicals and novel entities (such as insecticides)—for both of which the planetary boundaries are transgressed—form another important driver of population decline and subspecies extinctions for wild bees and domesticated bees alike [617]. The causal chain from specific operational decisions to apply biochemicals all the way to the extinctions of a particular beehive may again be difficult to establish. Nevertheless, the aggregated impact of a collection of operations management decisions to apply insecticides can be devastating and have considerable social and economic impact.

Bee population declines and extinctions may complicate access to food for humans as ecosystem services provided by wild bees and the services provided by domesticated bees pollinate more than 75 percent of the global food crop types.

The agricultural produce value at risk because of pollination loss is estimated to range between 235 and 577 billion USD annually [26, 151, 214]. These amounts are only indicative of financial consequences. Any "cost" estimate of the value of ecosystem services provided by bees is likely to be incorrect, as bee population decline and extinction across ecosystems cause cascading subsequent effects on plant biodiversity and other ecosystem services of which the net present financial value is beyond scientific understanding and may well be considered invaluable [139, 151].

We may recall from Chapter 9 that *bees as a service* are one of the existing operating models for pollination and might help to overcome shortages resulting from a reduction in ecosystem pollination services provided by wild bees. If the environmental impacts of existing agricultural operations continue to reduce bee populations, the food supply chains may even need pollination operations beyond those provided by bees. Substituting bees for humans has been estimated to be extremely costly, however, and robot bees (we may recall that the word drone originally refers to a male bee) are not yet (if ever) a viable and desirable alternative [457].

This example of the environmental impacts of biochemical flows and the creation of novel entities highlights the consequences operations management decisions may have for the planet and society, far beyond the costs and revenue of a single organization or value chain that typically form operations management objectives. Rather than taking up the impossible task of estimating these impacts and managing operations correspondingly, it may be more helpful in such cases to consider proxy objectives that more directly relate to operations. In relation to the example at hand, one may consider the process measure kilograms of biochemicals used. This approach can also be applied to phosphor and nitrogen, the two biochemicals for which the planetary boundaries have already been transgressed. More than one-third of supplementary phosphor and more than one-half of supplementary nitrogen are wasted and end up polluting terrestrial and coastal ecosystems, as illustrated in Figure 10.3 (and perhaps much more [166]).

For novel entities, a comprehensive set of measures along the various stages of the value chain that produce novel entities is, for instance, proposed in [443]. Many novel entities degrade slowly or not at all and accumulate in planetary ecosystems.

Today's food supply chains suffer from the inefficiencies of animal protein-based consumption patterns by humans at the top of the food chain and of wasting roughly one-third of the food produced [345]. From an equity perspective, it has been observed that these inefficiencies are an order of magnitude larger in high-income countries and occur while a population of 2 billion suffers from food deficiencies, many of which actually suffer from hunger (as addressed in SDG 2) [108]. Altogether, it appears that the environmentally problematic supplementary phosphor and synthetic nitrogen may not be needed if the operations in the corresponding value chains were managed to be lean and green. While the evidence of effective implementation still appears to be scarce, *green lean management* is indeed starting to receive attention both in agriculture and in manufacturing as an operations management practice that can help prevent transgressions of planetary boundaries [45, 215, 448].

While the primary sector still lags behind in harnessing the power of the 3rd industrial revolution, the 4th is already entering the sector and starting to contribute to its sustainability [346]. 4IR components, including AI, the IoT, and smart automation, are for instance being deployed in *precision agriculture*, irrigating and dosing nutrients as needed using data from sensors and about weather conditions [531]. Nanotechnologies that blend the 4IR technological advancements with the biological, for instance, in the form of nanomaterials operating within plants to improve crop yields, are in experimental stages of application [351]. The management of designing, implementing, and controlling such 4IR-based agricultural operations is in very early stages and mostly receives attention in terms of opportunities and challenges [346, 351, 531].

The planetary boundary regarding GHGs is directly related to operations management, as around 80 percent of these emissions are from the fossil fuels burned to operate the technologies of businesses. Especially the secondary sector, including the energy sector, plays a major role (see Table 1.1). The year 2022 marked an all-time high in industrial GHG emissions from fossil fuels [3]. The cumulative GHG emissions are directly and roughly linearly related to global warming (through radiative forcing) [356].

The urgent need to limit CO_2 emissions has been widely acknowledged in recent years, which in turn has resulted in increased attention for operations management methods and measures to reduce the carbon footprints of businesses and value chains. These methods include carbon footprinting, product life cycle assessment, and the Avoid-Shift-Improve (ASI) framework [70, 633]. Below, we present the RISA framework, which extends the ASI framework, to systematically cover operations management efforts to keep operations within the planetary CO_2 boundary. The RISA framework will play a central role in Chapter 11 as well.

The four ordered categories of operations management interventions to improve the sustainability of the **RISA framework** are:

Remove the products, waste, and other effects of operations that cause boundary transgressions,

Improve existing operations to reduce effects causing boundary transgressions,

Shift toward other operations that prevent or limit boundary transgressions, and

Avoid the operations causing boundary transgressions.

The RISA framework may be interpreted to present an order in which to consider sustainability improvements. For instance, if the negative effects on sustainability can be *removed*, doing so may be the simplest solution. It should be noted, however, that removal first requires to **capture** the pollutants and then to **reuse** or

store them. If removal is not possible, incremental changes might **improve** operations to avoid harmful effects. If improvement is infeasible or insufficient, a disruptive **shift** in operations may remedy their lack of sustainability. As long as none of these work, the solution is to **avoid** the operations. As we shall see in Chapter 11, solutions from all categories will be needed in a concerted effort to manage operations for a sustainable future. Below, we first apply the framework to reflect on current operations management practices regarding GHG emissions.

Emitted CO_2 can be *removed*, captured, and subsequently stored or reused. The original ecosystems of planet Earth are equipped with biotechnology to capture carbon from the atmosphere. Forestry operations management already contributes to expanding and enhancing the natural carbon capturing capacity of forests, which are known to function as **carbon sinks**. Novel technologies such as nanotechnology to enhance plant carbon capturing abilities, as discussed above, are in very early stages of deployment. With the exception of applications in fossil fuel production itself, fully mechanical carbon capturing technologies are also in the early stages of development and presently make a modest contribution to reducing GHG emissions [593]. Removal can also be applied to other GHGs, but it appears less feasible for various biochemicals and novel entities (such as microplastics).

CO_2 reductions in operations can next be achieved by *improving* energy efficiency. Technologies that reduce waste, e.g., by reducing the energy needed for production, avoiding overproduction, and not producing products with a short life cycle, are already beneficial. Such technological improvements can, for instance, be achieved by green lean operations management practices [45, 215, 448].

Many of the *shifts* in operations are based on shifting toward alternative emission-free energy resources. Renewable energy capacity increases are already growing much faster than fossil fuel-based capacity increases, and investments in renewable energy technologies are currently larger than investments in fossil fuel technologies [4].

Due to their virtually limitless availability, solar energy, wind energy, and ocean energy suffice to meet human demand for energy [593]. Current technologies using these energy sources typically produce electricity, and this yields operations management challenges in electricity networks because electricity storage technologies are still limited. Other existing technologies to replace fossil fuels include nuclear power production, geothermal-based energy production, and techniques that produce and use hydrogen.

The abundant availability of renewable energy resources and the cost improvements in their deployment, for instance, to produce electricity, are already shifting operations toward the use of green energy, such as electricity produced from renewable sources. This has increased operations management attention in the design and control of energy networks that accommodate the inflow of renewable energy produced and in energy storage within these networks (for instance, using new generations of lithium batteries and hydrogen). Moreover, trucks, buses, and cars increasingly operate on batteries charged with green electricity rather than the fossil-fueled combustion engines of the 2nd industrial revolution. Green hydrogen, i.e., hydrogen produced using renewable energy, can also be stored for later use in trucks and cars and in industrial processes (e.g., to produce "green steel" or "green cement") [479]. However, green hydrogen production and storage supply chains, which solely rely on sustainable energy, have not yet reached a stage of large-scale deployment [593].

From the above, it becomes clear how shifts in energy generation operations have already started to play an important role in reducing GHG emissions. The shift can also occur in primary operations powered by energy. For instance, the use of 3D printing can shift from classical subtractive manufacturing to additive manufacturing. The additive 3D printing operations avoid the waste that is typical of subtractive manufacturing, and their lean value creation facilitates many efficiency gains in the use of resources, including the use of energy resources and hence GHG emissions [296].

Lastly, *avoidance* of operations may entail the avoidance of intercontinental business travel (for which fossil-fueled aviation can be considered the only feasible alternative at the moment) or the avoidance of opening mines (or oil wells) as long as the negative impact on surrounding ecosystems from mining operations (e.g., regarding fresh water use or air pollution by novel entities) cannot be avoided.

Above, we reflected on current operations management practices in relation to the transgression of several of the most planetary boundaries, particularly those already transgressed or likely to be transgressed soon. From this overview, it appears that few of the technologies currently adopted to keep operations from causing transgressions of the planetary boundaries would be classified as 4IR technologies. However, some early-stage developments in nanobiology have, for instance, already been covered above, and the advancements in the use of AI to improve the efficiency of energy networks and in the use of 3D printing for sustainable construction are other noteworthy examples [296, 630]. As of now, however, the technologies adopted in operations management with the purpose of reducing GHG emissions are mostly pre-existing technologies rather than the technologies of the 4IR. As the 4IR is progressing dynamically and rapidly, the next chapter studies future contributions of 4IR technologies to steer human operations away from transgressions of the planetary boundaries.

DOI: 10.1561/9781638282853.ch11

Chapter 11

Operations for a Sustainable Future

Turn your face to the sun and the shadows fall behind you.

Maori proverb

11.1 Pathways Toward a Sustainable Future

The third research aim of this study is to explore how to redesign and manage human operations to form sustainable future ways of working and living for humankind, in particular by leveraging the technologies of the 4IR. While this may be viewed as the most relevant research aim, it is also the aim for which scientific methods tend to lack robustness and deliver answers with considerable uncertainties, with disclaimers, or with no answer at all. Data and evidence about the future are not available yet. Within these limitations, this chapter addresses the third research aim in the constructive and optimistic spirit of the indigenous wisdom cited in the opening lines. Thus, we build on the considerable scientific research into methods to mitigate transgressions of planetary and social boundaries and particularly aim to strengthen the understanding of their connection with operations and operations management.

The mitigation efforts often rely on the use of technology, including technology that still needs to be developed. We have already discussed how green hydrogen-based energy supply networks are still in their early stages, and the same holds for the uptake of nanotechnology to promote agricultural productivity. The somewhat loosely defined 4th industrial revolution is only beginning to unfold, while expected to fundamentally alter the operations of work and life. What plausible, scientific statements can be made about redesigning and managing future human operations without yet knowing the revolutionary technologies that may bring fundamental changes to how we operate in the near future?

To ensure the feasibility of our research aim, this subsection continues by excluding two possible yet unlikely scenarios of a truly disruptive nature. The next subsections then concretely explore future operations that are socially inclusive and within planetary boundaries, distinguishing agricultural, manufacturing, and service operations after starting with the operations of the energy sector [147, 410, 411, 575]. This structure differs from the structure of Chapter 10 yet better facilitates a solution-oriented perspective.

Perhaps the most revolutionary technology that has been suggested to resolve the challenges associated with transgressing the planetary boundaries is to start living and working outside of the planetary ecosystems. Of the options explored, moving part of humankind and their operations to planet Mars appears to be the most realistic [340, 554]. However, the ecosystems of Mars presently only contain a limited set of non-living elements and physical processes, while living organisms and products are fully absent.

For human life and operations on Mars, living organisms and products would have to be carried from planet Earth or manufactured on Mars. It is unlikely that any missions carrying even a single human being to Mars will happen before 2031. Moreover, the scientific literature does not offer any realistic possibilities for one-way travel and subsequent operations of human settlement on Mars [340, 554].

Even if such truly revolutionary developments happen on a small scale during the 21st century, there is presently no prospect of such operations outside of planetary ecosystems to arrive at a scale that could make a significant contribution to addressing the sustainability challenges for the growing human population of 8 billion humans and their operations on planet Earth. Hence, we disregard the pathway of resolving the challenges associated with the lack of sustainability of operations by locating (part of) the human population outside of the planetary boundaries.

Another truly revolutionary technological development would be for robots and other artificially intelligent devices to take (partial) control over operations presently controlled by humans. This could also happen in hybrid forms, where human biology blends with technology, as envisioned to be part of the 4th industrial revolution [144, 329, 517]. Many machines already have physical operational capabilities that go far beyond the capabilities of humans or other living organisms. What if machines might "*far exceed the human capacity for decision making in the real world*" in the foreseeable future [498]. Could they take over the control of the physical processes in ecosystems, i.e., could they take over control of operations and become the future managers of the operations of work and life?

The emergence of such machines has been predicted to become the "*biggest event in human history*" [498]. It will require to revise the definition of operations presented in Chapter 2, which assumes operations are controlled by living species. Experts, however, mostly position a (first) transition toward the emergence of machines that "*can carry out most human professions at least as well as a typical human*" several decades into the future [392].

Such a transition might form a stepping stone toward super-intelligent machines that can subsequently create even more intelligent machines, et cetera, and trigger further (revolutionary) developments impacting humankind and the planet [392, 498]. Such follow-up developments are mostly expected to take another number of decades rather than years [392].

Thus, for now, scientific consensus indicates that for the coming decade or two, humans will be in control of operations and be actively engaged in conducting these operations. As the transgression of planetary boundaries toward irreversible impacts on planetary ecosystems is importantly influenced by the operations of the advancements in the next two decades, we consider future scenarios of operations and operations management in which artificially intelligent machines take the lead to be out of scope for now. Doing so will not keep us from including the contributions of more readily available forms of AI to promote sustainable operations controlled by humans in the remainder.

The scientific analysis aimed at resolving the challenges associated with environmental and societal sustainability in general, and with regard to climate change in particular, importantly relies on a set of five commonly adopted "*shared socioeconomic pathways (SSPs)*" [147, 410, 411, 575]. These pathways are designed with reference to the socioeconomic challenges for mitigation and adaptation to climate change and explicitly relate to other dimensions of environmental sustainability, social inclusion, and economic development. For each of the four combinations of minor and major challenges for mitigation and adaptation, there is a corresponding pathway. A fifth pathway (SSP2) corresponds to medium challenges for both, as also illustrated in Figure 11.1.

The set of five pathways is not meant to form a complete set of exclusive pathways, one of which will underpin our actual future. They rather span a space of expected shared socioeconomic development possibilities as considered relevant for analysis of the sustainability of planet Earth and humankind until the end of the 21st century. Within this space, SSP2 may serve as a point of reference as it continues policy and business as usual.

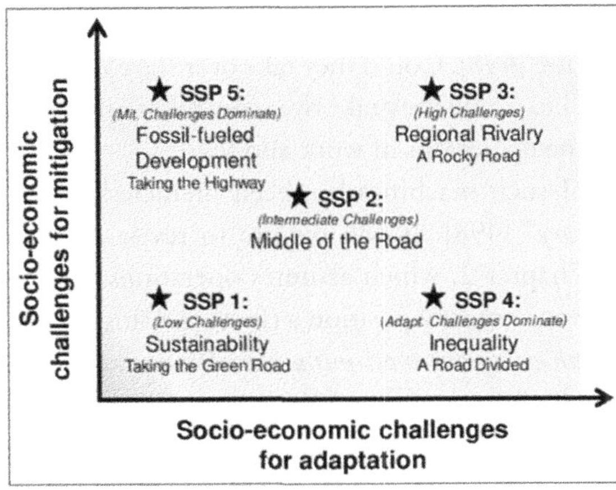

Figure 11.1. Five shared socioeconomic pathways and their associated challenges for mitigation and adaptation [411].

Below, we synthesize the five pathways based on the narratives provided in [410], with a special focus on operations and technological advancement.

SSP1: Sustainability—Taking the Green Road In this pathway "*The world shifts gradually, but pervasively, toward a more sustainable path, emphasizing more inclusive development that respects perceived environmental boundaries*" [410]. Inclusive well-being is prioritized over GDP growth, with corresponding access to healthcare and education as required for successful participation in the formal economy. Rapid technological change and the global spread of technological innovations enable the operations of work and life (including consumption) to become less resource and energy intensive and improve land use and environmental sustainability. As a result, this pathway entails relatively few challenges to mitigate climate change and improve social inclusion (in areas such as equitable access to food, education, and health services, income equality, and well-being) and implies reduced challenges to adaptation.

SSP2: Middle of the Road "*The world follows a path in which social, economic, and technological trends do not shift markedly from historical patterns. Development and income growth proceeds unevenly....*" This path is without technological breakthroughs, slows the spread of technological innovation to some global regions, and brings a decreasing but continued dependency on fossil fuels. Access to education remains inequitable and problematic for many. In combination with poverty and income inequalities in some regions, this limits social inclusion and particularly exposes the most vulnerable to climate change mitigation and adaptation challenges.

SSP3: Regional Rivalry—A Rocky Road "*Countries focus on achieving energy and food security goals within their own regions at the expense of broader-based development...Investments in education and technological development decline. Economic development is slow, consumption is material-intensive, and inequalities persist or worsen over time, especially in developing countries.*" The slow advancement and spread of agriculture technology and of technologies for cleaner production reflect the low priority attached to environmental sustainability. Altogether, this results in high challenges to mitigation and adaptation to climate change for many subpopulations across the globe.

SSP4: Inequality—A Road Divided "*Over time, a gap widens between an internationally connected society that is well educated and contributes to knowledge- and capital-intensive sectors of the global economy, and a fragmented collection of lower-income, poorly educated societies that work in a labor-intensive, low-tech economy.*" Technology advances and spreads rapidly in some sectors and mostly in developed countries; however, less educated subpopulations in these countries may experience less socioeconomic growth. Moreover, technological advancements in low-income countries are limited, which, together with limited access to education, yields a continued dependence on low-skilled, labor-intensive, informal operations, in particular in agriculture. This pathway is therefore lacking in many dimensions of social inclusion, such as income, poverty, access to water, food, and education. Any efforts made to improve environmental sustainability tend to be focused locally and in more prosperous regions. As resource-intensive operations are also concentrated in these regions, local measures might go a long way in keeping the global climate change mitigation challenges low. However, challenges to adaptation are high for the many vulnerable populations in poor socioeconomic conditions, confronted with local environmental challenges, or both.

SSP5: Fossil-Fueled Development—Taking the Highway "*Driven by the economic success of industrialized and emerging economies, the world places increasing faith in competitive markets, innovation, and participatory societies to produce rapid technological progress and development of human capital as the path to sustainable development....There are also strong investments in health, education, and institutions to enhance human and social capital. At the same time, the push for economic and social development is coupled with the exploitation of abundant fossil fuel resources and the adoption of resource- and energy-intensive lifestyles around the world.*" Effective advancement and spread of technology drives global productivity increases in operations and—along with equitable access to education, health services, and food—promotes social inclusion. Food production operations are effectively optimized, while the food consumption patterns of the global population present challenges for land use, among which is deforestation. More generally, this pathway entails high mitigation challenges and pressures on ecological boundaries, likely resulting in significant global warming. At the same time, socially inclusive economic progress enables adaptation challenges to be manageable for most humans.

Each of these pathways qualitatively describes a set of socioeconomic developments to underlie future scenarios for the environmental, social, and economic development of humankind on planet Earth. They provide input for quantitative scenario studies that integrate various environmental and socioeconomic dynamics and encompass corresponding policy measures to estimate the sustainability impacts associated with the pathways [478, 523]. More specifically, these studies may assess the likelihood with which each of the pathways brings futures within the environmental and social inclusiveness boundaries or causes transgressions of the boundaries of a safe and just operating space for the planet and humankind.

Initial studies already showed that mitigation policies to avoid transgressing the CO_2 boundaries were unlikely to exist for SSP3 and SSP5, nor for the Middle of the Road scenario for SSP2 [313, 478, 523]. Further and more detailed mitigation scenario studies confirm that continuing "*business as usual*," as represented by pathway SSP2, causes substantial transgressions. Current evidence suggests that SSP2-based scenarios limit global warming by the end of the century to 2.8 degrees at best [437, 522]. When also accounting for the differences between commitments made, policies subsequently put in place, and actual progress made, the temperature increase will exceed 3 degrees [437, 522]. We may subsequently recall from Chapter 10 that transgression of this boundary increases the risk of transgressing other ecological boundaries regarding biodiversity, freshwater use, and biochemical flows, and the risk of transgressing social inclusion boundaries regarding income, poverty, access to food, health, and others. Business as usual leads to a highly disruptive future for planetary ecosystems and human society.

More recent studies indicate that only taking the green road of SSP1 yields options to stay within the GHG boundaries of a safe operating space [522, 523]. As time proceeds, however, the disruptive changes in operations needed to stay within the GHG emissions boundaries are considered increasingly unlikely to be implemented on time. Much of the attention has therefore shifted to SSP1-based scenarios minimizing the temporary overshoot of the boundary and managing global warming to stay below 2 degrees [437, 522, 523].

As our interest is in sustainable operations, we disregard exploring SSP 2–5 in the remainder of the analysis and focus on operations and operations management practices in support of those scenarios for SSP1 that avoid or limit transgressions of the boundaries of a safe operating space. Given the uncertainties in the technological advancements of the current industrial revolution in operations and in the effectiveness of changes toward sustainable operations, the analysis will not look beyond 2050. Following SSP1, we therefore consider operations for a global population that grows to 8 billion in the coming decades [478]. Halfway the through century, more than 90 percent of this population is expected to be literate, and more than three-quarters of the population will be urbanized [478]. These scenarios also expect real average income to double from 2020 to 2050 and inequity as measured through the Gini coefficient to halve over the same period [147, 478].

11.2 Operations and Technology for a Sustainable Future

There are many interactions among the planetary boundaries, and hence there is no chain of logic that puts one of them before the others. Still, the GHG emissions boundary can be viewed as playing a central and pivotal role because of the pervasive influence of the climate changes that the net GHG emissions from human activity entail. In 2019, these net antrhopogenic emissions were higher than ever before and amounted to the equivalent of $59gTCO_2$. Almost two-third of the GHG emissions were caused by the use of fossil fuels to provide energy for human operations and by additional industrial emissions [437]. Total global GHG emissions from fossil fuels and industry were at an all-time high of 36 $GtCO_2$ in 2022 [347]. Scenarios that avoid or limit transgression of the GHG emissions boundary require a reduction to between 0 and 20 $GtCO_2$ by 2050 [437]. Commitments made by national governments until 2020—whether translated into policy or not—are forecast to close at most 20 percent of the gap to preventing GHG boundary transgressions until 2050 [437].

Energy systems will therefore be the first topic covered for the operations of a sustainable future. In addition to an in-depth exploration of the energy sector, this topic also covers energy-related changes in agriculture, forestry, land use, and the downstream secondary and tertiary sectors, up to and including the operations of life.

Agriculture, forestry, and land use drive most of the remaining GHG emissions. Agriculture, together with downstream operations in the food value chains, is a second area to be analyzed extensively. This analysis also zooms in on future operations to avoid (further) transgressions of other planetary boundaries, such as biochemical boundaries and social inclusion boundaries, for instance, in relation to SDG 2, zero hunger.

This section next turns to addressing remaining topics in the operations of work in industry (the secondary sector), services (the tertiary sector), and in the operations of life. Throughout, the analyses apply the RISA framework introduced in Chapter 10.

11.2.1 Energy and Greenhouse Gas Emissions

The energy needs of global operations may well grow in the coming decades as the global population grows, and the achievement of SDGs such as reduction of income inequities, elimination of poverty, access to housing, and access to clean energy will also entail increases in per capita energy use for the corresponding populations. Mitigation efforts needed to limit global warming to 2 degrees likely cause the total value produced by operations as reflected in global GDP to be 1 to 4 percent lower in 2050, in comparison to scenarios that exceed this global warming limit.

Methods from all four RISA categories are needed in mitigation scenarios associated with SSP1 to prevent or limit GHG emissions boundary transgression until 2050 [249]. For energy systems and the corresponding GHG emissions, the RISA framework can be restated more precisely as follows: **Remove** GHG emissions from operations; **Improve** energy systems to become more energy efficient; **Shift** toward energy sources that produce less GHG emissions; and **Avoid** high-emission energy production and products and services that require such energy.

A main avenue for the *removal* of GHG emissions is CO_2 removal, also known as carbon removal. A direct form of carbon removal is to add removal operations to the existing ones emitting these GHGs. **Carbon capturing and storing (CCS)** more generally refers to such direct capturing and subsequent storage of GHG emissions such as CO_2 and CH_4 from the production and use of fossil fuels. CCS can, for instance, reduce emissions from the production of combustible fuels and gases, plastics and other chemicals, and biofuels, and can in total make a contribution of 20 percent of the energy production-related GHG emission reductions required by effective mitigation scenarios for SSP1 [282].

If not captured directly, emitted GHGs can be removed from the atmosphere. Chapter 10 already mentioned how forests can function as carbon sinks. Reforestation and afforestation are removal operations included in all effective mitigation pathways. Pathways avoiding transgression of the GHG emission boundary may require an additional 322 million hectares of forest globally [437]. Forestry operations, which increase forest cover and improve the carbon capturing capacity of forests, are therefore important future operations. As discussed in Chapter 10, experimental advancements in nanotechnology may provide 4IR contributions to these operations. Oceans, together with the living species in the oceans, also form an important carbon sink.

In addition to ecosystem services to remove carbon from the atmosphere, man-made carbon removal technologies have been developed in recent decades. As of yet, these technologies are lacking in effectiveness and affordability to make a significant contribution to closing the GHG emissions gap [593].

As a final aspect of removal approaches, it is worth mentioning that captured carbon needs to be stored, as possible by ecosystem services, but can also be reused and form the input for subsequent operations, for instance as raw material for the production of plastics and building materials. These circular approaches transform value chains into value circles and can thus contribute beyond the capturing and removal itself [593].

Efficiency *improvement* of energy-producing operations and of operations powered by fossil fuels is the next RISA category to consider. Analytic and AI technologies from the 4IR realm are expected to make major contributions to improving the efficiency of operations in all sectors and of the energy systems feeding them [437]. Altogether, efficiency improvements might bring 25 percent of the required reduction in GHG emissions from energy production [282]. In the same 4IR realm, 3D printing and other forms of additive manufacturing may also contribute to efficiency improvement as they avoid the waste generated by subtractive manufacturing operations [296, 375].

The largest contribution to GHG emission reduction is envisioned to come from the third category in the RISA framework, *shift*. The importance of shifting away from fossil fuels is, for instance, illustrated by the emissions reductions from coal, oil, and gas of 100, 60, and 70 percent by 2050 in mitigation pathways that keep the planet within the GHG emissions boundary without carbon removal [437]. Less disruptive mitigation pathways that utilize carbon capturing and allow for a 2-degree temperature increase still require reduction percentages of, for instance, 85 (coal), 30 (oil), and 15 (gas) [437]. Variations in fossil fuel emissions across these three sources are possible when compensating increases in GHG emission reductions from one source or in one sector with GHG emission decreases elsewhere. These numbers demonstrate that all scenarios that prevent or limit GHG boundary transgressions require much more profound changes in energy production operations than those realized until 2022, when emissions from fossil fuels and industry were higher than before.

The required disruptive shift away from business as usual can largely be realized with the existing technologies presented in Chapter 10. A shift to using electricity as the (indirect) source of energy where and when possible avoids 20 percent of global GHG emissions. A shift away from fossil fuels to renewable energy resources, whether for electricity production or for direct use, yields another 25 percent and 35 percent when including green hydrogen [282].

All pathways that prevent or limit GHG boundary transgression rely on shifts toward emission-free electricity production (Green Electrification) in all sectors. In fact, nearly all electricity in pathways that are likely to limit global warming to 2 degrees is from low- or zero-carbon technologies using renewable resources such as the sun (e.g., photovoltaic systems), wind (turbines), and water [437]. The cost of energy produced by these renewable resources has decreased substantially recently and is now within the cost range of energy from fossil fuels [282, 437, 593].

For buildings, electrification implies that electricity provides the energy for cooking, heating, and air conditioning. Such a change in the energy system requires to build and strengthen existing electricity networks in some places—especially in developing countries—and to replace existing fossil fuel-based infrastructure (e.g., gas pipelines) in others. These applications for residential buildings and the availability of technologies in residential settings, such as solar cells and heat pumps, once again point at the mitigation opportunities for the operations of life. Capturing these opportunities often requires facilitating operations of work, for instance, to install and maintain the required technologies.

Zero-emission electricity can also form the energy source for most transportation on land when electrifying existing fossil fuel-powered fleets of motors, cars, trucks, trains, et cetera. A modality shift to walking or cycling is also helpful. Again, transformative shift is needed in the operations of life and in the operations of work.

Together, a shift toward green electricity and hydrogen can also replace the direct use of fossil fuels in most industrial operations. From an operations perspective, the shift toward zero-emission energy sources forms a shift toward a novel set of energy production and delivery operations, as well as a shift in the operation of the machines powered by new energy sources. This transition comes with additional advantages—such as the ubiquitous availability of the renewable resources sun and wind—and with disadvantages—such as the time and weather dependence of the availability of these resources. The operations management perspectives section below elaborates on the consequences of these shifts in energy provisioning for operations management, including the role of the 4IR to resolve the complex optimization challenges in the design and operation of these energy networks.

Extant technologies are insufficient to replace all direct use of fossil fuels as a source of energy by emission-free electricity or hydrogen, or to replace them at affordable cost levels [593]. For example, a shift toward green hydrogen to fuel airplanes and ships requires further technological advancement. More generally, the operation of affordable and effective green hydrogen supply chains, including production, transport, storage, and delivery, still requires further technological advancement before practical use at scales that make a significant contribution to GHG emission reduction [437]. Biofuels, produced from plants, are an additional alternative, yet they come with the complexity of competing with agriculture and forests for scarce land [522].

Electricity can also be produced by nuclear energy. As recent progress in renewable alternatives has been much more effective, nuclear energy plays a modest role in most mitigation pathways [478].

The small reduction in GDP growth associated with scenarios for SSP1, compared to other scenarios, hints at the contribution *avoidance* can make as a mitigation instrument. Avoidance will be necessary for some of the products and services for which Removal, Improvement, and Shift presently cannot reduce GHG emissions as required. This can apply to fossil fuel-based transportation modalities, in particular air travel and sea travel, until shifts to emission-free alternatives are available. Avoidance can also take the form of avoiding GHG intensive raw building materials such as cement and steel, which is particularly feasible when constructing smaller buildings. The operations of life offer many additional opportunities for avoidance, in particular in relation to reducing purchases of new goods such as electronic devices, clothes, and cars that, at a population level, have a large carbon footprint. Transforming value chains into value circles that reuse and recycle goods can additionally enable the avoidance of GHG emissions.

Altogether, avoidance plays an important and significant role. The main mitigation efforts in sustainable pathways, however, are not based on avoiding the automated operations of the industrial revolutions but rather on shifting away from the use of fossil fuels to power them. The two Shankey diagrams in Figures 11.2 and 11.3 illustrate an energy transition toward sustainable operations corresponding to net zero emissions from 2060 onward. They illustrate a reduction in final energy use

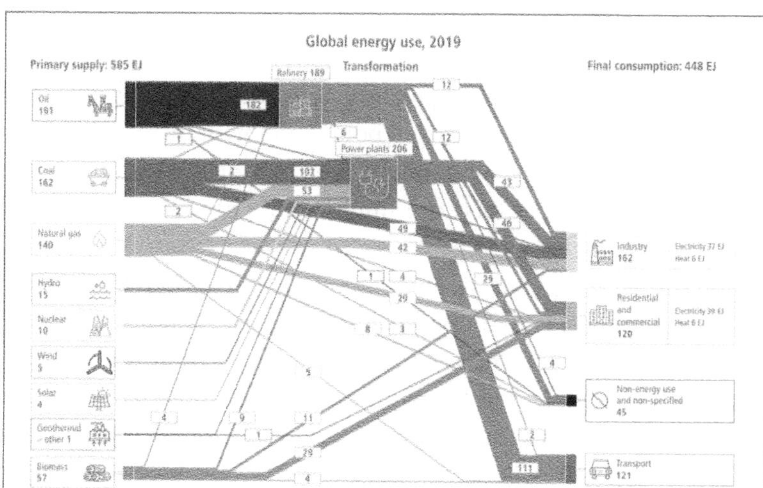

Figure 11.2. Shankey diagram of 2019 Global Energy Flows (reprinted from Figure 6.1 (Panels: Global energy use, 2019 and Global energy use, 2060, Scenario IMP-REN-2.0 [117]).

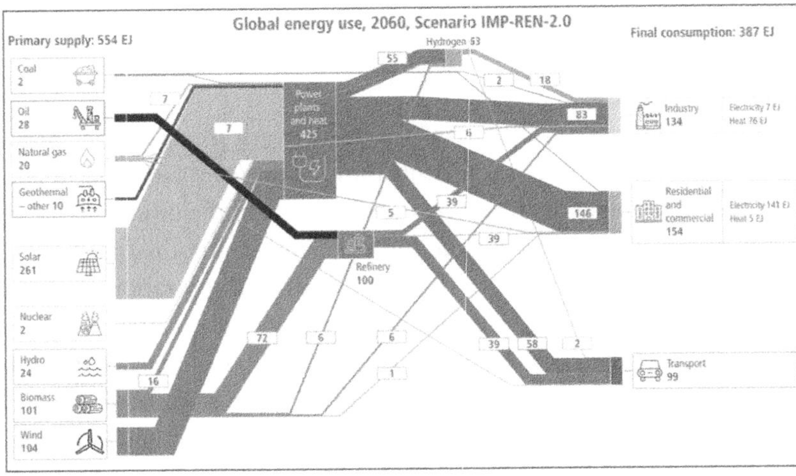

Figure 11.3. Shankey diagram of 2060 Global Energy Flows for a renewable energy-based scenario that reaches net zero GHG emissions by 2060 (reprinted from Figure 6.1 (Panels: Global energy use, 2019 and Global energy use, 2060, Scenario IMP-REN-2.0 [117]).

of around 15 percent, while shifting toward renewable energy reduces fossil fuel use by more than 85 percent. Other scenarios with zero net anthropogenic emissions require lower end-user energy demand (avoid) or more removal [117].

11.2.2 Agriculture, Land Use, and Biochemical Flows

As described above and in Chapter 10, land use can promote sustainability through reforestation and afforestation operations, creating carbon sinks that remove CO_2 from the atmosphere. Conversely, deforestation to free up land for the production of food, cotton, and other materials increases the likelihood of transgressing the GHG emissions boundary. When operating agriculture as usual, the food needs of a growing and more prosperous global population with increased animal source food in their diets—require a substantial larger part of the surface of the planet to be used for agriculture and food consumption and production as usual implies deforestation.

Mitigation pathways, by contrast, incorporate disruptive and substantial changes in agricultural operations, food supply chains, and diets, as outlined below. In these mitigation pathways, the operations of agriculture, forestation, and other forms of land use contribute 20 percent or more of the GHG emissions reductions achieved in mitigation pathways that avoid or limit transgressing the corresponding boundary [437]. Biofuels play a delicate mitigation role as land used to grow crops for biofuels cannot be forested. Biofuels thus reduce both CO_2 emissions and CO_2 capturing. A net-effectiveness analysis of biofuels that takes alternatives into account thus requires careful assessment.

Agriculture impacts many sustainability dimensions other than the GHG emissions boundary. As is clear from Chapter 10, operations for a sustainable future need to resolve their problematic ecological impact, as they have already transgressed the biochemical flows and biodiversity boundaries. The same chapter also explained how agriculture is a key determinant of the social inclusion dimension of access to food. Advancements in agricultural operations co-determine the accomplishment of SDGs 2 and 3: zero hunger and health and well-being for all. While the percentage of the global workforce employed in agriculture has decreased, it still employs more than one in four, many of whom are small-scale farmers. Future agricultural operations are therefore also of importance to meet social inclusion goals related to the eradication of poverty (SDG 1), decent work (SDG 8), and income equity (SDG 10).

This subsection again adopts the RISA framework to explore sustainable future agricultural operations in relation to other forms of land use and food value chains.

The *removal* of agricultural biochemicals from soil and water, for instance, needed to preserve terrestrial ecosystems as well as freshwater ecosystems and coastal ecoystems, mostly relies on ecosystem services. Technological advancements to remove these biochemicals appear to be limited at present. The same applies to technologies for large-scale GHG capturing in agriculture. There are, however, valuable contributions to be made by capturing and using food waste itself. Roughly one-third of human food produced globally is wasted [247]. Moreover, agricultural GHG emissions are importantly caused by the GHGs, such as CH_4, emitted by food waste that ends up in landfills.

As evidenced by agricultural practices since the onset of the agricultural revolution, removed food waste can serve as biofertilizer, thus improving the circularity of agriculture (see also below) [558]. Removed food waste can also serve as feedstock to produce biofuels and biochar, thus forming a circular substitute for fossil fuels [226].

Current technologies already offer various opportunities for the **improvement** of the efficiency of agricultural operations. On the one hand, these improvement opportunities involve better and possibly more intensive use of farmlands that currently provide modest yields and whose soil may become less fertile because of a lack of water and fertilizers. This holds especially true for farmlands in low- and middle-income countries whose operations have not fully adopted the innovations of preceding agricultural revolutions that have taken place elsewhere in recent centuries.

These efficiency improvements are important as 2 billion people live from small-scale farming, and there is little evidence that they are already significantly benefiting from technological advancement (see also Chapter 10) [49, 187, 369]. Their context and operations make them especially vulnerable to global warming, drought, and other forms of climate change. Advancements in operations are likely needed to prevent a negative impact on agricultural yield through nutrition depletion, soil erosion, and desertification [384]. Efficiency improvements from technological advancements may combine access to affordable inputs such as water, phosphor, and nitrogen with access to (artificially intelligent) agriculture 4.0 technologies, such as precision agriculture [531].

On the other hand, precision agriculture and other innovations are needed to reduce fertilizer inputs in many high-income countries and some rapidly developing countries. In these contexts, more efficient use can contribute to undoing transgressions of planetary biochemical boundaries. Chapter 10 already reported that more than one-third of supplementary phosphor and more than one-half of supplementary nitrogen (and perhaps much more) are wasted. The uptake of *green lean management* practices in agricultural operations management, presently in the early stages of adoption, can substantially reduce these inefficiencies [45]. Using sensors, precision agriculture technology can assess the fertilization needs of a variety of nutrients and subsequently deliver (precisely) the quantities needed using controlled-release fertilizer based on nanotechnology or other recent biotechnological advancements [166]. Similar technologies can improve the efficiency of the use of pesticides and insecticides, thus reducing the risks of (local) biodiversity boundary transgressions.

Other forms of efficiency improvement in agriculture, for instance, include ruminant diet improvements that reduce their CH_4 emissions [239, 364]. Efficiency improvements in human food procurement and diets can also serve to mitigate sustainability risks. Diets that include more calories than necessary can cause obesity while adversely affecting planetary ecosystems [16, 360]. Obesity has, for instance, been estimated to cause 1.6 percent of global GHG emissions, thus indicating a possible area for efficiency improvement for the operations of life.

Mitigation efforts that *shift* toward alternative food production operations can also make substantial contributions to (returning to) staying within planetary boundaries. The use of agricultural land to produce animal-based foods for humans consumes more than three-quarters of global agricultural land [514]. A shift to diets that are more plant-based and provide plant-based proteins therefore frees up land for forests and avoids the GHGs emitted by livestock such as CH_4. Moreover, it reduces the use of fertilizers, especially nitrogen, which is an essential component of proteins, and thus contributes to repairing the transgression of the biochemical flow boundary. This shift again tightly links sustainable production and consumption (SDG 12).

A shift away from land-based agriculture is another possibility. Existing alternatives include cellular fermentation, cultured meat, controlled environment agriculture (greenhouses occupying land not suitable for agriculture), and soilless agriculture (for instance, vertical agriculture in urban environments). Recent technological innovations, including those from the 4th industrial revolution, for instance in nanobiology and 3D printing of food, can enable and promote this shift toward alternative food production operations [437].

A shift in the fertilizers used, replacing (manufactured) nitrogen and phosphor with biomass and recently developed nanofertilizers, is another change in agricultural operations that can contribute to containing the transgression of the biochemical flow boundaries [89, 110].

Just as it is not necessary to undo the industrial revolutions to mitigate GHG emission boundary transgression, sustainable food production does not necessitate to undo the agricultural revolution and return to hunting and gathering. Still, opportunities for the *avoidance* of agricultural operations exist as follows from the aforementioned evidence: around one-third of the present food produced for humans is wasted or lost [247, 628] . Moreover, 35 to 40 percent of GHG emissions from food are associated with wasted and lost food, accounting for 8 to 10 percent of global GHG emissions [437, 628]. Lean agriculture, which avoids waste, can therefore make a very substantial contribution to reducing GHG emissions.

Obviously, avoiding the production of food that is subsequently wasted or lost also helps the mitigate transgression of the biochemical boundary (for nitrogen and phosphor) and the freshwater boundary. The total area of land used to produce human food that ends up being wasted is roughly four times larger than the area required for reforestation in mitigation scenarios for SSP1 [247, 437]. Thus, avoiding the production of wasted food yields opportunities for carbon removal and biodiversity. Similar observations can be made for other agricultural products such as cotton. Cotton forms 25 percent of the base materials for clothing. Global clothing consumption and production have more than doubled since the turn of the millennium, and the shortening of product life cycles has already caused the annual volumes of clothing waste ending up in landfills to grow to 100 million tons [403] (see Figure 11.4).

11.2.3 Industry and Novel Entities

The energy systems of the secondary sector, industry, as well as industry's contribution to food value chains have already been addressed above. Here we turn to the remaining sustainability topics of future manufacturing operations with a special view toward the advancements of Industry 4.0 and beyond.

Industrial operations produce plastics, metals, and other novel entities that advance downstream the global supply chains yet ultimately are not consumed by the global population but remain in the planetary ecosystems after having contributed to the operations of life. The byproducts of industrial operations add

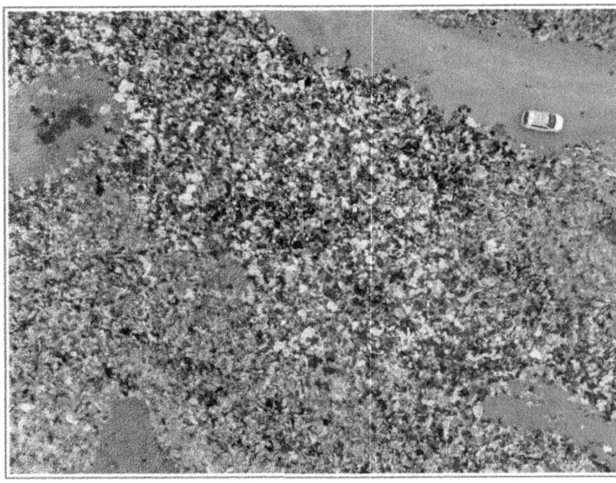

Figure 11.4. Unsold global fast fashion dumped in the Chilean Atacama desert (source Martin Bernetti/AFP via Getty Images).

further to the plastics, metals, novel entities, and chemicals that enter planetary ecosystems. As we have seen in Chapter 10, these material flows already cause transgressions of the novel entity boundary. In addition, they negatively impact human health and biodiversity, resulting in further negative impacts on terrestrial and ocean ecosystems.

Removal of end-of-life industrial products with their plastic and metal components, rubber, batteries, et cetera, and subsequent reuse (recycling) when possible as a material for next operations can play a key role in preventing pollution of ecosystems and staying within local and planetary biochemical and novel entity boundaries. This requires the design and management of **reverse supply chains**, i.e., networks of entities whose combined operations collect, transport, and recover or sustainably store disposed products (including byproducts) [573]. While hampered by a variety of barriers, among which workforce and management capacity may be among the most important, manufacturers are increasingly developing such reverse supply chains [319, 573], which are also called value cycles. These value cycles often necessarily develop together with public service operations, in particular waste collection, to avoid dumping consumer products in landfills and the environmentally harmful combustion of waste [573].

Future waste reverse supply chains can improve on existing practices of bringing waste to locations where disposal and burning practices damage the environment and the health of local human populations and other living species, as discussed in Chapter 10 [7, 583]. Improvements are first needed for the population of 2 billion living without formal waste collection services and to protect the health and well-being of the "*11 million informal entrepreneurs who work closely with waste, delivering a circular economy but often without protective equipment or a structured, safe system of work*" [583]. Proper waste management, especially of electronic waste, is needed to prevent diseases such as cancer and respiratory and cardiovascular diseases [7, 469].

Plastic forms the majority of waste entering oceans annually, and continuing plastic business and consumption as usual will cause the oceans to contain more plastic than fish by 2050 [102]. Globally, up to 40 percent of plastic waste from coastal communities may end up in the ocean, as many of these communities lack effective public waste collection services [596]. The annual value of plastic packaging materials disposed into the oceans alone may amount to 80 billion 2015USD globally [102, 486]. It is causing much damage to ocean ecosystems, some of which are already beyond repair and thus form a transgression of the planetary novel entity boundary [486, 587].

The prevention of further damage to ocean ecosystems needs plastic removal operations beyond waste collection in the communities of disposal. Such removal is difficult for microplastics but feasible for macroplastics some of which can be recycled (or upcycled) to produce shoes, clothes, bottles, sunglasses, skateboards, et cetera [154, 596]. More than a hundred enterprises have business models centered around ocean plastic prevention, removal, and recycling operations. Few, if any, of these have progressed beyond prototyping and scaled up operations toward volumes of removal that approach the order of magnitude of accumulated ocean plastic [154]. The discussion on ocean waste will continue below under the improvement and avoidance categories of the RISA framework.

Improvement of the efficiency of industrial operations offers a variety of opportunities to strengthen their sustainability. Let us consider some of the most relevant opportunities, other than the energy-efficient improvements already addressed above. Upstream in the manufacturing supply chains, there are possibilities to reduce material inputs and the use of non-sustainable materials. This can, for instance, be achieved by (additive) manufacturing techniques (in particular, 3D printing) to produce right-sized components. Such Industry 4.0 manufacturing operations may take the form of prefabricated components or postpone manufacturing to on-site production. For concrete production, such material efficiencies additionally bring emission reductions of 24 to 50 percent [437].

The importance of plastics and the growing volume of plastic production are another area where efficiency improvements are possible. Many of these opportunities start with the reduced use of plastics in downstream applications. More efficient packaging, for instance, translates into lower demands for manufacturing upstream and less plastic waste downstream. This applies to packaging materials and consumer products alike.

Short product life cycles imply a higher ratio of materials and manufacturing operations over product use. Similar observations have been made above in relation

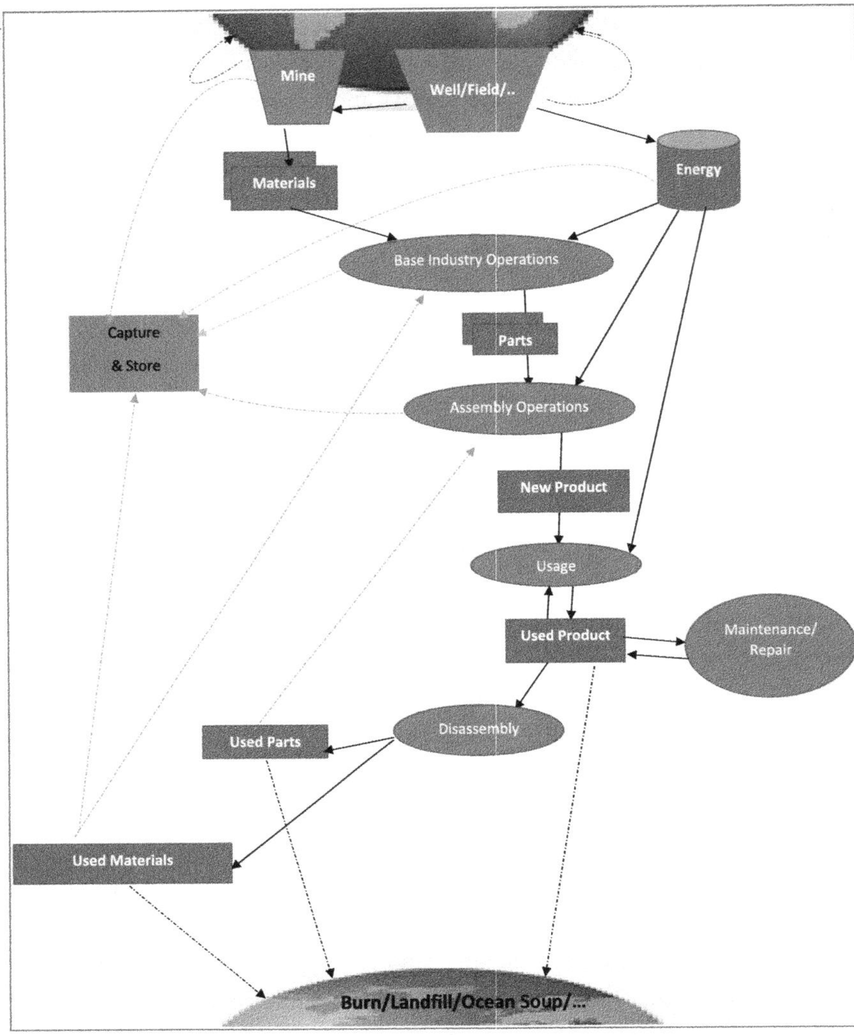

Figure 11.5. A full material life cycle view of manufacturing supply chains, adapted from [352]. Lines refer to material flows. Black dashed lines refer to pollution; green-dashed lines refer to removal and subsequent storage or reuse.

to fast fashion, which may also include non-organic materials such as nylon. These life cycle assessment considerations, which should include recycling, call for *cradle to grave* and *cradle to cradle* approaches to the assessment of environmental footprints [474]. Figure 11.5 provides an elaboration of the extended value networks that may be considered for manufacturing operations, which include recycling (cradle to cradle), storage, and waste (cradle to grave) flows.

The use of completely automated, lights-out, manufacturing, in which the operations planning and control are optimized using APS software, can maximize efficiency as much as possible and incorporate related sustainability measures as specified. Many hardware and software solutions are already available, and hence the dissemination of these practices is an important priority for future operations and operations management.

Chapter 10 discussed how the automated nature of these operations of the 4th industrial revolution implies a reduction in the need for lowly skilled, low-wage labor. Lights-out manufacturing reduces this need to zero. This has implications for the social inclusion of the low-skilled workforce, especially in countries with low wages, which are currently attractive manufacturing locations in global manufacturing supply chains. The advancements of the 4IR likely imply that location decisions for manufacturing operations will be reconsidered and more closely aligned with the availability of a highly skilled workforce to manage these advanced automated systems and with the markets of end consumers they serve. These developments are already part of the business cases of present reshoring and backshoring projects [441]. Such location decisions may more completely consider the sustainability of the supply chain, including access to clean energy, water, possibilities for waste disposal, et cetera.

While reshoring may thus promote the environmental footprints of global manufacturing supply chains, the effect on social inclusion may be negative. They may reduce incomes for lowly skilled manufacturing workers and their households and also negatively impact other social inclusion dimensions, among which poverty, decent work, access to food, housing, and healthcare for these populations. From a social inclusion perspective, these developments in future operations need to be complemented by creating alternative operations of work for those negatively impacted, preferably based on education and upskilling toward a more highly skilled workforce.

Shifts toward alternative manufacturing operations can also make important sustainability contributions. From the above, it is also already clear that shifting toward recycled and recyclable materials can bring substantial benefits. This may, for instance, entail to substitute cementitious material with ground limestone and calcined clays and to replace metals and plastics with (bio)degradable polymers and organic materials [251, 437]. Shifting away from metals and plastics reduces the short- and long-term environmental damage caused by microplastics and nanoplastics, toxic metal nanoparticles, and other novel entities, even though negative environmental and health impacts may not yet be fully understood [251, 295]. The role of plastics is, however, ambivalent as plastics can also replace heavier and less efficiently produced materials (e.g., metals).

Avoidance of manufacturing at large is unlikely to happen, as the expected growth of the global population and their wealth likely drives up consumption and hence production of manufactured goods until 2050. Business as usual therefore implies a continued increase in industrial production, products, their use, and subsequent waste [549]. This overall increasing trend, however, enlarges opportunities for avoidance of certain manufactured goods and manufacturing operations and materials, in the spirit of the global GDP increase from operations projected for SSP1 that is slightly lower by 2050 than in alternative pathways [437].

Some of the opportunities for avoidance are closely related to the aforementioned measures from other RISA categories. Avoiding transportation services to avoid their GHG emissions may also entail avoiding manufacturing the corresponding airplanes and vehicles. Likewise, resource sharing schemes, such as car sharing schemes, may result in a reduction in car production. Avoiding the production of wasted food also avoids packaging materials, et cetera. Additionally, avoidance can come from prolonging product life cycles, or at least slowing down the current trends of shortening product life cycles that bring along shorter life cycles of manufacturing resources as well [352, 549].

11.2.4 Service Operations

The service sector with the largest impact on environmental sustainability is the transportation sector. Its impact mostly relates to the combustion of fossil fuels and resulting GHG emissions as already covered above. Another important service sector contribution to GHG emissions, stems from the energy used for heating and cooling buildings. These energy needs will be impacted by the ongoing climate changes, with increased needs for cooling in some places and reduced needs for heating in other places or in other seasons. Any GHG emission impacts from heating and cooling need to be mitigated by green electrification of heating and cooling systems, as covered above. These two mitigation interventions from the Shift category can cover much of the sustainability risks associated with large and expanding service sectors such as the hospitality and leisure sector and the health services sector. These service sectors can further make contributions by improving the sustainability of food services, in particular the reduction of food waste, along the lines already discussed above.

Chapter 10 revealed that despite the stellar growth in production and use of information and communication technology of the 3rd industrial revolution, the energy consumed by operating these technologies to provide information and communication services is in the order of 1 percent of global energy demand and increases only modestly as technology adoption advances rapidly because of simultaneous efficiency improvements (see also [437]).

All in all, the RISA framework serves to present solutions for the environmental challenges of the service sector. The situation is more complex for the social inclusion challenges, and even the SSP1-based mitigation pathways fail to achieve all SDGs. The difficulties are severest for the SDGs 1, 2, and 4, which aim to eradicate poverty, achieve zero hunger, and ensure secondary education for all [134, 636]. The latter is especially relevant for service operations. We have seen in the preceding chapters that the technological stagnancy of the service sector caused it to provide a place to work for the large majority of the workforce in developed countries, and this development can be expected to occur in developing countries for SSP1-based scenarios as the primary and secondary sectors become more technology-intensive and productive and drive income growth. The same 4th industrial revolution driving these advances in the primary and secondary sectors, however, may also disrupt the service sector and diminish its technological stagnance. This reduces the capacity of the service sector to provide work opportunities and, hence, to absorb workforce inflow from other sectors.

We have seen in Chapter 10 that with the ongoing advances in AI, service (ro)bots and other technologies increasingly replace humans in service operations as their cognitive and relational capabilities increase. Following common business case logic, return on investment for these technologies benefits from large-scale application, as might be attainable when substituting larger volumes of relatively lowly skilled service workers. The technology thus may first place lowly skilled service workers at risk of losing (decent) work and income. Over the next decades, business cases for the substitution of higher-skilled workers can be expected to become more abundant as well, as AI machines become more capable and versatile. While there is still much uncertainty around these developments regarding the future of work, AI may, for instance, reduce employment opportunities for skilled workers in financial services (accountancy, tax services, and insurance), medical services (radiology, home health services, and prevention), and hospitality and leisure services. The first robot restaurants and hotels already exist [].

Like it has been the case in preceding industrial revolutions, one may expect that the transformation will provide some job opportunities. Ford's assembly line required more than 200,000 relatively low-skilled workers between the mechanical machines, and AT&T once employed more than half a million switchboard operators. Are the delivery workers and warehouse employees working between the robots in the huge distribution centers of the online economy a 4IR equivalent, with comparable difficulties of access to decent work and income? For these workers, an inequitable low salary and hence low income may (temporarily) protect them from being substituted by robots. This reminds us that the 4IR may diminish social inclusion.

For high-income workers, the small scale of their operations may importantly cause the return on investment of technology to replace them to be insufficient. Moreover, the unfolding of the 4IR will provide job opportunities for high-skilled workers developing technologies and new operating systems and implementing these systems, standing in the shoes of James Hargreaves, Lilian Gilbreth, and

Frederick Brooks. Recent evidence, however, suggests that the number of new jobs created by recent technological advancements is modest [55, 208].

If and when the 4th industrial revolution causes the primary and secondary sector to operate with less human labor input and the tertiary service sector loses important components of its capacity to absorb this human labor capacity because of the same 4th industrial revolution, the future of the operations of work becomes unclear. How can the future operations of work provide decent work (SDG 8) and serve a larger, better educated (SDG 4), and more prosperous (SDG 1) human population within the environmental and social boundaries as the 4IR advances? Will salary from the operations of work continue its decline as a source of income for the global population? This may sound attractive, yet it may threaten income and income equality for lowly skilled workers and for those who lack other sources of income. Solutions to the challenges to social inclusion posed by the 4IR may therefore additionally need to come from mechanisms beyond market mechanisms to determine salaries, from finding large-scale alternatives to salary as a source of income, or both.

The social inclusion challenges associated with the advancement of the 4IR in the service sector are of a different nature than the often environmentally driven challenges within the primary and secondary sectors. Their impact is, as of yet, uncertain but may be equally important. This is especially noteworthy as technologies, methods, and frameworks to address these social inclusion challenges appear relatively less developed.

11.3 Operations Management Reflections for Sustainable Operations Management

Whereas preceding chapters revealed how operations management has played its part in creating sustainability challenges, this chapter shows how it can make essential, meaningful contributions to resolving the sustainability challenges [70]. Operations management activities can be devoted to (re)designing and implementing sustainable future operations of work and life as defined in Chapter 2. Given the large contribution of business to boundary transgressions, the pivotal role operations management can play in the transformation toward sustainable operations especially belongs to businesses and brings a prime responsibility for operations management as a key area of business management. Business operations management decisions have been and are "*principle contributors to anthropogenic effects on ecosystem sustainability*" [163], and there is much work ahead to improve their sustainability impacts, as they have recently started to receive significant attention from the operations management domain [591].

The International Panel for Climate Change duly recognizes the role of operations management, for instance, when stating that "*integrated energy planning and operations that take into account energy demand and system constraints across all sectors, ..., offer the opportunity to leverage sectoral synergies and avoid inefficient allocation of energy resources*" [437]. Several key aspects of operations, such as material efficiency, emerging primary process innovations, value cycles that reduce waste, and other innovations from Industry 4.0, have, however, not yet been explicitly modeled in the integrated assessment models used to develop sustainability pathways [437]. Narrowing the gap between the operations management community and the sustainability community may help craft feasible and effective pathways toward sustainable operations.

For the business operations management discipline, a repositioning of the domain toward sustainability implies a new set of values and competences. It requires to primarily value environmental protection and social inclusion, where the latter includes economic aspects of operations including employee incomes, stakeholder profits, and the costs of products and services for consumers. Life cycle assessment, carbon foot printing (Shankey diagrams), waste management, and management of water, biochemical flows, and novel entities will be valuable core competencies of future operations managers. In addition, operations need to be designed to equitably provide decent work and income.

As indicated by the RISA framework, an extant focus on continuous improvement, as advocated in 6 σ and lean management, can make a partial, valuable contribution in the transition toward sustainable business operations. This especially applies to the challenges that can be resolved by incorporating sustainability into existing continuous improvement frameworks, such as in green lean management [215]. Manufacturing companies that have adopted a differentiation and innovation-based operations strategy have already tended to show effective prioritization of ecological and social sustainability alongside improved business performance [350]. Such incremental changes to existing operations are especially effective when the context provided by government and society is supportive [603].

However, the I for Improvement, i.e., incremental changes to business as usual (including pledges already made), will not suffice to stay within the boundaries of a safe and just operating space [437]. Moreover, the valuable efforts toward circular operations—representing the R(emove) of the RISA framework—have thus far only resulted in minor contributions to achieving sustainability goals [437]. Disruptive advancements in operations from the RISA categories S for Shift and A for Avoid will need to make major contributions. Business operations management therefore needs to emphasize disruptive improvement methods over, or in addition to, continuous improvement approaches.

The shift from fossil fuels to renewable resources in energy value chains can enable further shifts in many operations. This shift can bring fundamental redesigns as the sources of energy value chains shift from fossil fuel rich locations to places where renewable energy sources such as sun, wind, and water are abundant and can be affordably transformed into electricity and fuels.

For instance, small-scale renewable energy production technologies and operations (e.g., solar cells) that can be deployed close to the point of use for the operations of work and life have already appeared to be efficient and flexible and have diffused relatively fast in recent years [437]. These developments disrupt the operations of the energy sector, reducing the need for remote production and increasing the need to provide networks that support electrified, shorter, locally centered value networks in the energy sector.

In locations providing affordable and sustainable access to water and renewable energy resources, newly designed (or redesigned) energy networks need to accommodate alignment with the production of zero- or low-emission fuels such as green hydrogen (or ammonia). More so, proximity to such locations may also become an important factor in industrial facility location, especially for energy-intensive manufacturing operations requiring (fuel-based) combustion and steel manufacturing operations [437].

The changes in energy and manufacturing value network designs and locations will be additionally affected by the expected changes in population sizes and income, which cause changes in the distribution of global demand for energy and manufactured goods. The implementation of SSP1-based mitigation scenarios will elevate the importance of sustainable operations and supply chain management in general and of the energy sector in particular for several decades to come.

In a way, the developments may be most disruptive for the populations and organizations in countries whose economies have been significantly industrialized since the first industrial revolution of the late 18th century. Organizations and citizens of these countries need to unlearn and change ways of working and living, possibly even destroying previously created value, as is the case when stranding assets for unsustainable operations.

Less industrialized countries have invested less in unsustainable technologies and operations, which offers opportunities to achieve decent work and income, access to food, housing, and energy, and other social inclusion targets without having to avoid or shift existing operations, to strand assets, or to unlearn corresponding operations management practices. However, the transition toward sustainability may be more difficult in less industrialized countries as their operations management capabilities are generally less developed, particularly for sustainable operations [88]. Together with other barriers posed, such as (lack of enabling) governmental legislation and policies and access to financial resources, this may hamper the effective and timely implementation of transformation projects [319]. Building sustainable operations management skills in low and lower-middle income countries, which represent more than half of the global population, will thus be a key success factor for global sustainability [514].

The spread of innovations across borders and cultures to these areas, including sustainable operations management practices, has not been a core strength or priority in operations management in the past. In fact, business as usual is often based on protective measures to prevent the spread of innovations through patents and property rights. Such models sharply contradict the advancements in technology transfer and capacity building needed to meet climate objectives and more broadly contribute to sustainable development [437]. Disruptive steps forward in the spread of operations management and capacity building are called for, even when adopting a realistic perspective on the achievement of sustainability targets and SDGs [437].

The transition may also be particularly difficult in regions whose operations and value creation heavily depend on the supply of fossil fuels. More so, the relative ease of value creation in the energy sector may have caused operations and operations management in other sectors to have advanced less. A reduction in value create on from fossil fuels then needs to be accompanied by developing operations and operations management skills in other sectors to create value and avoid loss of societal wealth and social inclusion. If fossil fuel energy exports reduce and are not replaced by renewable energy exports, imports of goods and services from other sectors may need to reduce correspondingly. This requires to develop operations and local supply chains to better cover local demand by setting up sustainable and competitive local production and service operations.

The management of service operations will be a delicate societal matter across the globe in the coming decades. Industry 4.0 and Agriculture 4.0 can be expected to scale up and bring technological advancements that further substitute, especially low-skilled, human labor with technology. Thus, the 4th industrial revolution may further add to the inflow of human labor into the service sector. However, AI machines of the 4th industrial revolution may add further momentum to the service revolution that is starting to resolve the long-standing technological stagnance of the service sector and enable further substitution of human labor by technology in low- and high-skilled service jobs [55, 277].

It is too early to predict the effects of the service revolution on global employment. However, the trends indicate that new jobs do not arise in the same locations where extant jobs become obsolete. If current evidence is indicative of the future, new opportunities may not appear as much in new jobs as in upskilled versions of extant ones [55]. From an operations management perspective, this calls for capturing the opportunities provided by the 4th industrial revolution to redesign services and service operations so as to provide decent work and income, especially for those at risk of losing decent work and income because of the same revolution. If, over time, technological advancements cause global value creation to become less human labor intensive, it may be good to recall from Chapter 9 that extant hunter-gatherers frequently allocate less than 8 hours a day to the operations of work [42, 581].

The thus defined challenge for service operations management implies to avoid repeating the pathways of previous industrial revolutions that adopted operating models that provide lowly skilled and lowly paid work for women and men operating between machines. Sustainable operations management prioritizes decent, socially inclusive work over efficiency, cost, and profit.

The same sustainability perspective also applies to the operations of life, implying a shift toward responsible consumption and away from a subset of utilitarian values and hedonic values of happiness and well-being that underlie the unsustainable consumption and production targeted in SDG 12 [6]. Different values of utility and well-being can motivate more sustainable consumption, for instance, purchasing sustainably produced clothes rather than fast fashion that generates waste and yields non-inclusive low incomes for the workers in the value chain [269]. The RISA framework can also facilitate a transformation toward sustainable consumption and, more generally, sustainable operations of life. Sustainable choices in the operations of life occur downstream of the value cycles and chains in which businesses and other organizations can thus form a very powerful purchasing lever toward sustainable human operations. The urgent need to bring the operations of life into the focus of operations that need to be managed and disrupted is further emphasized by the minor sustainability contributions innovations toward circular operations have made thus far [437].

Lack of evidence of positive impacts from individual transformations in the operations of life, together with the lack of an affirmative and enabling context (e.g., for waste collection and recycling), may diminish the belief that individual contributions matter and can be perceived as demotivating. Evidence of successful impacts by innovators and early adopters is therefore important, as it can promote a wider adoption of sustainable practices [489]. Moreover, innovators, early adopters, and anyone else can bring their values and beliefs to their operations of work, whether they work as operations managers, in operations, or otherwise. After all, businesses, governmental organizations, NGOs, and all other organizations are human-created structures for the operations of work and/or life, and the intrinsic motivations of all involved are a most effective force to direct operations management efforts toward operating safely and justly within the planetary and social boundaries [335, 603].

References

[1] Abulhab, S. D. (2017). *The Law Code of Hammurabi: Transliterated and Literally Translated from its Early Classical Arabic Language*. Blautopf Publishing.

[2] Administration of the International Energy Agency. "Website". URL: https://www.iea.org/reports/pulp-and-paper (accessed on 09/21/2022).

[3] Administration of the International Energy Agency. "Website". URL: https://www.iea.org/reports/co2-emissions-in-2022 (accessed on 06/01/2023).

[4] Administration of the International Energy Agency. "Website". URL: https://www.iea.org/reports/renewable-energy-market-update-june-2023/executive-summary (accessed on 06/01/2023).

[5] Administration, Federal Highway. "State motor vehicle registrations by year". URL: https://www.fhwa.dot.gov/ohim/summary95/mv200.pdf (accessed on 02/09/2021).

[6] Adomaviciute, K. (2013). "Relationship between utilitarian and hedonic consumer behavior and socially responsible consumption". *Economics and management*. 18(4): 754–760.

[7] Akram, R., S. Fahad, M. Z. Hashmi, A. Wahid, M. Adnan, M. Mubeen, N. Khan, M. I. A. Rehmani, M. Awais, M. Abbas, *et al.* (2019). "Trends of electronic waste pollution and its impact on the global environment and ecosystem". *Environmental Science and Pollution Research*. 26: 16923–16938.

[8] Allen, R. C. (2000). "Economic structure and agricultural productivity in Europe, 1300–1800". *European Review of Economic History*. 4(1): 1–25.

[9] Allen, R. C. (2009). *The British industrial revolution in global perspective*. Cambridge University Press.

[10] Allsopp, M. H., W. J. De Lange, and R. Veldtman. (2008). "Valuing insect pollination services with cost of replacement". *PloS one*. 3(9): e3128.

[11] Alperson-Afil, N. (2008). "Continual fire-making by hominins at Gesher Benot Ya 'aqov, Israel". *Quaternary Science Reviews*. 27(17–18): 1733–1739.

[12] Alvarez, R. A. and E. Paranhos. (2012). "Air pollution issues associated". *Air Pollution Issues Associated. EM Feature June.*

[13] Ambrose, S. H. (2001). "Paleolithic technology and human evolution". *Science.* 291(5509): 1748–1753.

[14] America – Model T Ford Club. "Encyclopedia Model B". URL: https://www.mtfca.com/encyclo/earlyfds.htm (accessed on 05/09/2021).

[15] America – Model T Ford Club. "Factory Layout". URL: https://www.mtfca.com/books/15-factory/ (accessed on 05/09/2021).

[16] An, R., M. Ji, and S. Zhang. (2018). "Global warming and obesity: a systematic review". *Obesity Reviews.* 19(2): 150–163.

[17] Anand, S. and A. Sen. (1994). "Human Development Index: Methodology and Measurement."

[18] Andersen, H., K. A. Røvik, and T. Ingebrigtsen. (2014). "Lean thinking in hospitals: is there a cure for the absence of evidence? A systematic review of reviews". *BMJ open.* 4(1): e003873.

[19] Anderson, C. B., G. M. Pastur, M. V. Lencinas, P. K. Wallem, M. C. Moorman, and A. D. Rosemond. (2009). "Do introduced North American beavers Castor canadensis engineer differently in southern South America? An overview with implications for restoration". *Mammal Review.* 39(1): 33–52.

[20] Antras, P. (2015). *Global production: Firms, contracts, and trade structure.* Princeton University Press.

[21] Arcas, B. A. y and A. Fairhall. (2001). "Archaeology of type". *Nature.* 411(6841): 997–997.

[22] Archives, IBM Archives. "IBM 360 – Form Computers to Computer Systems". URL: https://www.ibm.com/ibm/history/ibm100/us/en/icons/system360/impacts/ (accessed on 02/01/2023).

[23] Arms, R. G. (1959). "From Disassembly to Assembly". *Cincinnati: The Birthplace of Mass.*

[24] Assembly – UN General Assembly. (1948). "Universal declaration of human rights". *UN General Assembly.* 302(2): 14–25.

[25] Atasu, A., C. J. Corbett, X. Huang, and L. B. Toktay. (2020). "Sustainable operations management through the perspective of manufacturing & service operations management". *Manufacturing & service operations management.* 22(1): 146–157.

[26] Atkins, J. and B. Atkins. (2017). *The business of bees: An integrated approach to bee decline and corporate responsibility.* Routledge.

[27] Atkinson, A. B., E. Marlier, and B. Nolan. (2004). "Indicators and targets for social inclusion in the European Union". *JCMS: Journal of Common Market Studies.* 42(1): 47–75.

[28] Attwell, L., K. Kovarovic, and J. R. Kendal. (2015). "Fire in the Plio-Pleistocene: the functions of hominin fire use, and the mechanistic, developmental and evolutionary consequences". *Journal of anthropological sciences*. 93: 1–20.

[29] Augarten, S. (1984). *Bit by Bit: An Illustrated History of Computers*. Houghton Mifflin Co.

[30] Austin, H. A. (1928). "History and development of the beet sugar industry". *PhD thesis*. Colorado State University. Libraries.

[31] Bacco, M., P. Barsocchi, E. Ferro, A. Gotta, and M. Ruggeri. (2019). "The digitisation of agriculture: a survey of research activities on smart farming". *Array*. 3: 100009.

[32] Bailey, R. (1994). "The other side of slavery: Black labor, cotton, and textile industrialization in Great Britain and the United States". *Agricultural History*. 68(2): 35–50.

[33] Bailey, V. (1926). "How beavers build their houses". *Journal of Mammalogy*. 7(1): 41–44.

[34] Bairoch, P. and J.-M. Limbor. (1968). "Changes in the industrial distribution of the world labour force, by region, 1880–1960". *Int'l Lab. Rev.* 98: 311.

[35] Bank – The World Bank. (2023a). "Tourism and Competitiveness". URL: https://www.worldbank.org/en/topic/competitiveness/brief/tourism-and-competitiveness.

[36] Bank – The World Bank. "Poverty and Inequality Platform". URL: https://pip.worldbank.org/home (accessed on 08/04/2023).

[37] Bank – The World Bank. "World Development Indicators". URL: https://databank.worldbank.org/metadataglossary/gender-statistics/series/NY.GDP.MKTP.CD (accessed on 11/09/2021).

[38] Bank – The World Bank. "Data". URL: data.worldbank.org (accessed on 10/08/2021).

[39] Bar-Yosef, O. (1986). "The walls of Jericho: an alternative interpretation". *Current Anthropology*. 27(2): 157–162.

[40] Barber, R. M., N. Fullman, R. J. Sorensen, T. Bollyky, M. McKee, E. Nolte, A. A. Abajobir, K. H. Abate, C. Abbafati, K. M. Abbas, *et al.* (2017). "Healthcare Access and Quality Index based on mortality from causes amenable to personal health care in 195 countries and territories, 1990–2015: a novel analysis from the Global Burden of Disease Study 2015". *The lancet*. 390(10091): 231–266.

[41] Barbieri, P., A. Boffelli, S. Elia, L. Fratocchi, and M. Kalchschmidt. (2022). "How does Industry 4.0 affect international exposure? The interplay between firm innovation and home-country policies in post-offshoring relocation decisions". *International Business Review*. 31(4): 101992.

[42] Barnard, A. (1983). "Contemporary hunter-gatherers: Current theoretical issues in ecology and social organization". *Annual review of anthropology*. 12(1): 193–214.

[43] Baron, J. H. (2009). "Sailors' scurvy before and after James Lind–a reassessment". *Nutrition reviews*. 67(6): 315–332.

[44] Barragán, R. (2017). "Working silver for the world: mining labor and popular economy in colonial Potosí". *Hispanic American Historical Review*. 97(2): 193–222.

[45] Barth, H. and M. Melin. (2018). "A Green Lean approach to global competition and climate change in the agricultural sector–A Swedish case study". *Journal of cleaner production*. 204: 183–192.

[46] Baten, J. and J. L. Van Zanden. (2008). "Book production and the onset of modern economic growth". *Journal of Economic Growth*. 13(3): 217–235.

[47] Baumol, W. J. (1967). "Macroeconomics of unbalanced growth: the anatomy of urban crisis". *The American economic review*. 57(3): 415–426.

[48] Baumol, W. J., S. A. B. Blackman, and E. N. Wolff. (1985). "Unbalanced growth revisited: asymptotic stagnancy and new evidence". *The American Economic Review*: 806–817.

[49] Baumüller, H. (2018). "The little we know: an exploratory literature review on the utility of mobile phone-enabled services for smallholder farmers". *Journal of International Development*. 30(1): 134–154.

[50] Becerra-Valdivia, L. and T. Higham. (2020). "The timing and effect of the earliest human arrivals in North America". *Nature*. 584(7819): 93–97.

[51] Beck, K., M. Beedle, A. Van Bennekum, A. Cockburn, W. Cunningham, M. Fowler, J. Grenning, J. Highsmith, A. Hunt, R. Jeffries, *et al.* (2001). "Manifesto for agile software development".

[52] Behar, D. M., R. Villems, H. Soodyall, J. Blue-Smith, L. Pereira, E. Metspalu, R. Scozzari, H. Makkan, S. Tzur, D. Comas, *et al.* (2008). "The dawn of human matrilineal diversity". *The American Journal of Human Genetics*. 82(5): 1130–1140.

[53] Bellwood, P. (2004). *First Farmers: The Origins of Agricultural Societies*. Wiley. ISBN: 9780631205661. URL: https://books.google.fr/books?id=k1mzyCEOAUIC.

[54] Benito Alvarez, J.-M. "Oldowan Traditional Chopper". *Wikimedia Commons*. URL: https://commons.wikimedia.org/wiki/File:Oldowan_tradition_chopper.jpg (accessed on 08/12/2023).

[55] Berger, T. and C. B. Frey. (2016). "Structural transformation in the OECD: Digitalisation, deindustrialisation and the future of work".

[56] Berners-Lee, M. (2022). *The carbon footprint of everything*. Greystone Books Ltd.

[57] Best, E. L. (1933). *The change from manual to dial operation in the telephone industry*. No. 110. US Government Printing Office.

[58] Bettinger, R. L., L. Barton, and C. Morgan. (2010). "The origins of food production in north China: A different kind of agricultural revolution". *Evolutionary Anthropology: Issues, News, and Reviews: Issues, News, and Reviews*. 19(1): 9–21.

[59] Bilgen, S. (2016). "The environmental effects of coal-related activities". *Energy Sources, Part A: Recovery, Utilization, and Environmental Effects*. 38(22): 3283–3289.

[60] Black, B. and M. Ladson. (2010). "Oil at 150: Energy past and future in Pennsylvania". *Pennsylvania Legacies*. 10(1): 6–13.

[61] Black, J. (2011). *A brief history of slavery: A new global history*. Hachette UK.

[62] Blakey, R. C. (2008). "Gondwana paleogeography from assembly to breakup—A 500 my odyssey". *Geological Society of America Special Papers*. 441: 1–28.

[63] Bloom, D. E., D. Canning, R. Kotschy, K. Prettner, and J. J. Schünemann. (2019). "Health and economic growth: reconciling the micro and macro evidence". *Tech. rep.* National Bureau of Economic Research.

[64] Bocquet-Appel, J.-P. (2011). "The agricultural demographic transition during and after the agriculture inventions". *Current Anthropology*. 52(S4): S497–S510.

[65] Bolt, J., M. Timmer, and J. V. Luiten. (2014). "GDP per capita since 1820".

[66] Bonan, G. B. (1999). "Frost followed the plow: Impacts of deforestation on the climate of the United States". *Ecological Applications*. 9(4): 1305–1315.

[67] Bonnet, F., J. Vanek, and M. Chen. (2019). "Women and men in the informal economy: A statistical brief". *International Labour Office, Geneva*. 20.

[68] Boomgaard, P. (2017). "The forests of southeast Asia, forest transition theory and the Anthropocene, 1500–2000". *Economic Development and Environmental History in the Anthropocene: Perspectives on Asia and Africa*. p. 179.

[69] Boström, M., A. M. Jönsson, S. Lockie, A. P. Mol, and P. Oosterveer. (2015). "Sustainable and responsible supply chain governance: challenges and opportunities". *Journal of Cleaner Production*. 107: 1–7.

[70] Bouchery, Y., C. J. Corbett, J. C. Fransoo, and T. Tan. (2016). *Sustainable supply chains: A research-based textbook on operations and strategy*. Vol. 4. Springer.

[71] Bourguignon, F. and C. Morrisson. (2002). "Inequality among world citizens: 1820–1992". *American economic review*. 92(4): 727–744.

[72] Bowles, S. and J.-K. Choi. (2019). "The Neolithic agricultural revolution and the origins of private property". *Journal of Political Economy*. 127(5): 2186–2228.

[73] Braveman, P., S. Egerter, and D. R. Williams. (2011). "The social determinants of health: coming of age". *Annual review of public health*. 32: 381–398.

[74] Braveman, P. and S. Gruskin. (2003). "Defining equity in health". *Journal of Epidemiology & Community Health*. 57(4): 254–258.

[75] Brazier, R. E., A. Puttock, H. A. Graham, R. E. Auster, K. H. Davies, and C. M. Brown. (2021). "Beaver: Nature's ecosystem engineers". *Wiley Interdisciplinary Reviews: Water*. 8(1): e1494.

[76] Broadberry, S., H. Guan, and D. D. Li. (2018). "China, Europe, and the great divergence: a study in historical national accounting, 980–1850". *The Journal of Economic History*. 78(4): 955–1000.

[77] Broadberry, S. and B. Gupta. (2009). "Lancashire, India, and shifting competitive advantage in cotton textiles, 1700–1850: The neglected role of factor prices 1". *The Economic History Review*. 62(2): 279–305.

[78] Broadberry, S. and B. van Leeuwen. (2008). "British economic growth and the business cycle, 1700–1850: annual estimates". *University of Warwick*.

[79] Brooks Jr, F. P. (1995). *The mythical man-month: essays on software engineering*. Pearson Education.

[80] Brosnan, S. F. (2011). "Property in nonhuman primates". *New directions for child and adolescent development*. 2011(132): 9–22.

[81] Brown, D. K., A. Deardorff, and R. Stern. (2004). "The effects of multinational production on wages and working conditions in developing countries". In: *Challenges to globalization: Analyzing the economics*. University of Chicago Press. 279–330.

[82] Brown, K. W. (2012). *A history of mining in Latin America: from the colonial era to the present*. UNM Press.

[83] Brown, M. T., G. Protano, and S. Ulgiati. (2011). "Assessing geobiosphere work of generating global reserves of coal, crude oil, and natural gas". *Ecological Modelling*. 222(3): 879–887.

[84] Bulliet, R. W. (2016). *The wheel: Inventions and reinventions*. Columbia University Press.

[85] Bureau – United States Census Bureau. "Historical Estimates of World Population". URL: https://www.census.gov/data/tables/time-series/demo/international-programs/historical-est-worldpop.html (accessed on 11/09/2021).

[86] Buringh, E. and J. L. Van Zanden. (2009). "Charting the "Rise of the West": Manuscripts and Printed Books in Europe, a long-term Perspective from the Sixth through Eighteenth Centuries". *The Journal of Economic History*. 69(2): 409–445.

[87] Bush, V. (1931). "The differential analyzer. A new machine for solving differential equations". *Journal of the Franklin Institute*. 212(4): 447–488.

[88] Caiado, R. G. G., O. L. G. Quelhas, D. L. d. M. Nascimento, R. Anholon, and W. Leal Filho. (2019). "Towards sustainability by aligning operational programmes and sustainable performance measures". *Production Planning & Control.* 30(5–6): 413–425.

[89] Calabi-Floody, M., J. Medina, C. Rumpel, L. M. Condron, M. Hernandez, M. Dumont, and M. de La Luz Mora. (2018). "Smart fertilizers as a strategy for sustainable agriculture". *Advances in agronomy.* 147: 119–157.

[90] Canadell, J. G. and M. R. Raupach. (2008). "Managing forests for climate change mitigation". *Science.* 320(5882): 1456–1457.

[91] Cardozo, E. S., J. B. F. Araújo Neto, A. Barza, A. C. C. França, and F. Q. da Silva. (2010). "SCRUM and productivity in software projects: a systematic literature review". In: *14th International Conference on Evaluation and Assessment in Software Engineering (EASE).* 1–4.

[92] Carlen, J. (2016). *A Brief History of Entrepreneurship.* Columbia University Press.

[93] Carmi, E. (2015). "Taming Noisy Women: Bell Telephone". female switchboard operators as a noise source". *Media history.* 21(3): 313–327.

[94] Carpenter, S. R. and E. M. Bennett. (2011). "Reconsideration of the planetary boundary for phosphorus". *Environmental Research Letters.* 6(1): 014009.

[95] Cashdan, E. A. (1980). "Egalitarianism among hunters and gatherers". *American Anthropologist.* 82(1): 116–120.

[96] Census – United States Census Bureau. (1952). *Census of Population, 1950: Industrial Characteristics Special Report.* US Government Printing Office.

[97] Census – United States Census Bureau. (1966). *Long Term Economic Growth, 1860–1965: A Statistical Compendium.* US Department of Commerce, Bureau of the Census.

[98] Census – United States Census Bureau. (1973a). *Census of Population, 1970, Supplementary Report, Detailed Industry of Employed Persons by Race and Sex for the United States: 1970.* US Government Printing Office.

[99] Census – United States Census Bureau. (1973b). *Census of Population, 1970, Supplementary Report, Detailed Occupation of Employed Persons by Race and Sex for the United States: 1970.* US Government Printing Office.

[100] Census – United States Census Bureau and Social Science Research Council (US). (1949). *Historical statistics of the United States, 1789–1945: a supplement to the Statistical abstract of the United States.* US Government Printing Office.

[101] Census – United States Census Bureau. (1960). *Historical statistics of the United States: Colonial times to 1957.*

[102] Centre, United Nationes News Centre. (2017). "UN's mission to keep plastics out of oceans and marine life". URL: https://www.un.org/sustainabledevelopment/blog/2017/04/feature-uns-mission-to-keep-plastics-out-of-oceans-and-marine-life (accessed on 07/08/2023).

[103] Chancel, L., T. Piketty, E. Saez, and G. Zucman. (2022). *World inequality report 2022*. Harvard University Press.

[104] Chapin III, F. S., P. A. Matson, and P. Vitousek. (2011). *Principles of terrestrial ecosystem ecology*. Springer Science & Business Media.

[105] Chaplin, C. "Modern Times". URL: https:photo.charlychaplin.com (accessed on 01/02/2023).

[106] Chartier, Y. (2014). *Safe management of wastes from health-care activities*. World Health Organization.

[107] Chazan, M. (2017). "Toward a long prehistory of fire". *Current Anthropology*. 58(S16): S351–S359.

[108] Chen, C., A. Chaudhary, and A. Mathys. (2020). "Nutritional and environmental losses embedded in global food waste". *Resources, Conservation and Recycling*. 160: 104912.

[109] Chepeliev, M., M. Maliszewska, I. Osorio-Rodarte, S. E. Pereira, M. Filipa, and D. Van Der Mensbrugghe. (2022). "Pandemic, Climate Mitigation, and Reshoring".

[110] Chew, K. W., S. R. Chia, H.-W. Yen, S. Nomanbhay, Y.-C. Ho, and P. L. Show. (2019). "Transformation of biomass waste into sustainable organic fertilizers". *Sustainability*. 11(8): 2266.

[111] Chondros, T. G. (2009). "The development of machine design as a science from classical times to modern era". In: *International Symposium on History of Machines and Mechanisms: Proceedings of HMM 2008*. Springer. 59–68.

[112] Chondros, T. G. (2010). "Archimedes life works and machines". *Mechanism and Machine Theory*. 45(11): 1766–1775.

[113] Church, R. (1989). "Production, employment and labour productivity in the British coalfields, 1830–1913: some reinterpretations". *Business History*. 31(3): 6–27.

[114] Churchman, C. W. (1970). "Operations research as a profession". *Management science*. 17(2): B–37.

[115] Cioffi-Revilla, C. (1996). "Origins and evolution of war and politics". *International Studies Quarterly*. 40(1): 1–22.

[116] Cipolla, C. M. (1972). "The diffusion of innovations in early modern Europe". *Comparative studies in Society and History*. 14(1): 46–52.

[117] Clarke, L., Y.-M. Wei, A. De La Vega Navarro, A. Garg, A. Hahmann, S. Khennas, I. Azevedo, A. Löschel, A. Singh, L. Steg, G. Strbac, K. Wada, and (eds.) (2023). "Energy Systems". In: *Climate Change 2022: Mitigation*

of Climate Change. Contribution of Working Group III to the Sixth Assessment Report of the Intergovernmental Panel on Climate Change. Cambridge University Press, Cambridge, UK and New York, NY, USA.

[118] Cobb, J. and H. Daly. (1989). "For the common good, redirecting the economy toward community, the environment and a sustainable future".

[119] Coe, L. (1995). *The telephone and its several inventors: A history.* McFarland.

[120] Conceicao, P. (2022a). "Technical Notes Human Development Report 2021-22". URL: https://hdr.undp.org/sites/default/files/2021-22_HDR/hdr2021-22_technical_notes.pdf.

[121] Conceicao, P. (2022b). "Uncertain times, unsettled lives: shaping our future in a transforming world". URL: https://hdr.undp.org/system/files/documents/global-report-document/hdr2021-22pdf_1.pdf.

[122] Connan, J., A. Nissenbaum, and D. Dessort. (1992). "Molecular archaeology: Export of Dead Sea asphalt to Canaan and Egypt in the Chalcolithic-Early Bronze Age (4th-3rd millennium BC)". *Geochimica et Cosmochimica Acta.* 56(7): 2743–2759.

[123] Connors, J., J. Gwartney, and H. Montesinos-Yufa. (2020). "The rise and fall of worldwide income inequality, 1820–2035". *Southern Economic Journal.* 87(1): 216–244.

[124] Conservation of Nature, International Union for the Conservation of Nature. "Red List Summary Statistics 2021". URL: https://www.iucnredlist.org/resources/summary-statistics (accessed on 09/19/2021).

[125] Cook, D. and B. Davíðsdóttir. (2021). "An appraisal of interlinkages between macro-economic indicators of economic well-being and the sustainable development goals". *Ecological Economics.* 184: 106996.

[126] Copeland, B. J. (2004). *The essential turing.* Clarendon Press.

[127] Corrigan, J. M. *et al.* (2005). "Crossing the quality chasm". *Building a better delivery system.* 89.

[128] Cortada, J. W. (2019). *IBM: The rise and fall and reinvention of a global icon.* MIT Press.

[129] Coşgel, M. M., T. J. Miceli, and J. Rubin. (2012). "The political economy of mass printing: Legitimacy and technological change in the Ottoman Empire". *Journal of Comparative Economics.* 40(3): 357–371.

[130] Cotta, J., M. Breque, L. De Nul, and A. Petridis. "Industry 5.0". URL: https://research-and-innovation.ec.europa.eu/knowledge-publications-tools-and-data/publications/all-publications/industry-50-transformative-vision-europe_en (accessed on 03/15/2023).

[131] Coulson, M. (2012). *The history of mining: The events, technology and people involved in the industry that forged the modern world.* Harriman House Limited.

[132] Cox, S. (2002). "Information technology: the global key to precision agriculture and sustainability". *Computers and electronics in agriculture*. 36(2–3): 93–111.

[133] Creasy, R. J. (1981). "The origin of the VM/370 time-sharing system". *IBM Journal of Research and Development*. 25(5): 483–0–490.

[134] Crespo Cuaresma, J., W. Fengler, H. Kharas, K. Bekhtiar, M. Brottrager, and M. Hofer. (2018). "Will the Sustainable Development Goals be fulfilled? Assessing present and future global poverty". *Palgrave Communications*. 4(1).

[135] Crisman, P., G. Schroeder, and L. Pouzin. (1964). "Programming Staff Note 39".

[136] Current, R. N. (1951). "Technology and Promotion: The Typewriter". *Business History Review*. 25(2): 77–83.

[137] D'Andreamatteo, A., L. Ianni, F. Lega, and M. Sargiacomo. (2015). "Lean in healthcare: a comprehensive review". *Health policy*. 119(9): 1197–1209.

[138] Daily, G. (2003). "What are ecosystem services". *Global environmental challenges for the twenty-first century: Resources, consumption and sustainable solutions*: 227–231.

[139] Daily, G. C. *et al.* (1997). "Introduction: what are ecosystem services". *Nature's services: Societal dependence on natural ecosystems*. 1(1).

[140] Darwin, C. (2016). "On the origin of species, 1859".

[141] David, P. A. (1985). "Clio and the Economics of QWERTY". *The American economic review*. 75(2): 332–337.

[142] Davies, T. J., B. H. Daru, B. S. Bezeng, T. Charles-Dominique, G. P. Hempson, R. M. Kabongo, O. Maurin, A. M. Muasya, M. van Der Bank, and W. J. Bond. (2020). "Savanna tree evolutionary ages inform the reconstruction of the paleoenvironment of our hominin ancestors". *Scientific Reports*. 10(1): 1–8.

[143] Davies, W. and P. Gollop. (2003). "The human presence in europe during the Last Glacial Period II. Climate tolerance and climate preferences of Mid-and Late Glacial hominids". *Neanderthals and modern humans in the European landscape of the last glaciation*: 131–146.

[144] Davis, N. *et al.* (2016). "What is the fourth industrial revolution". In: *World Economic Forum*. Vol. 19.

[145] De Haan, H. (2009). "Building the Great Pyramid by Levering. A Mathematical Model". *PalArch's Journal of Archaeology of Egypt – Egyptology*. 6(2): 01–22.

[146] De La Vega, E., T. B. Chalk, P. A. Wilson, R. P. Bysani, and G. L. Foster. (2020). "Atmospheric CO2 during the Mid-Piacenzian Warm Period and the M2 glaciation". *Scientific Reports*. 10(1): 1–8.

[147] Dellink, R., J. Chateau, E. Lanzi, and B. Magné (2017). "Long-term economic growth projections in the Shared Socioeconomic Pathways". *Global Environmental Change*. 42: 200–214.

[148] DeMenocal, P. B. (2001). "Cultural responses to climate change during the late Holocene". *Science*. 292(5517): 667–673.

[149] Department, Reserve Bank of India Statistics Department. "The India KLEMS Database". URL: https://rbi.org.in/Scripts/KLEMS.aspx (accessed on 01/23/2023).

[150] Department, Statistics Division of the Department of Economic and Social Affairs, United Nations. "System of National Accounts 2008, 2008 SNA". URL: https://unstats.un.org/unsd/nationalaccount/sna2008.asp (accessed on 01/10/2023).

[151] Díaz, S. M., J. Settele, E. Brondízio, H. Ngo, M. Guèze, J. Agard, A. Arneth, P. Balvanera, K. Brauman, S. Butchart, *et al.* (2019). "The global assessment report on biodiversity and ecosystem services: Summary for policy makers".

[152] Dickinson, G. and K. Murphy. (2007). *Ecosystems*. Routledge.

[153] Dickinson, H. W. (2011). *A short history of the steam engine*. Cambridge University Press.

[154] Dijkstra, H., P. van Beukering, and R. Brouwer. (2021). "In the business of dirty oceans: Overview of startups and entrepreneurs managing marine plastic". *Marine Pollution Bulletin*. 162: 111880.

[155] Disease Control – Centers for Disease Control and Prevention. (2021). "National Health Expenditures". URL: https://www.cdc.gov/nchs/data/hus/2020-2021/HExpType.pdf.

[156] Disease Control – Centers for Disease Control and Prevention. "Health, United States – Data Finder". URL: https://www.cdc.gov/nchs/data/hus/2020-2021/HExpGDP.pdf (accessed on 01/05/2023).

[157] Division – The United Nations Statistics Division. "Basic Set of Environment Statistics". URL: https://unstats.un.org/unsd/envstats/fdes/basicset.cshtml (accessed on 12/09/2021).

[158] Donaghy, K., G. Rudinger, and S. Poppelreuter. (2004). "Societal trends, mobility behaviour and sustainable transport in Europe and North America". *Transport Reviews*. 24(6): 679–690.

[159] Donaldson, M. S., J. M. Corrigan, L. T. Kohn, *et al.* (2000). "To err is human: building a safer health system".

[160] Dondi, C. (2013). "The European printing revolution". *The Book, A Global History*: 80–92.

[161] Dorigo, M. and C. Blum. (2005). "Ant colony optimization theory: A survey". *Theoretical computer science*. 344(2–3): 243–278.

[162] Dorigo, M. and L. M. Gambardella. (1997). "Ant colony system: a cooperative learning approach to the traveling salesman problem". *IEEE Transactions on evolutionary computation*. 1(1): 53–66.

[163] Drake, D. F. and S. Spinler. (2013). "OM forum—Sustainable operations management: An enduring stream or a passing fancy?" *Manufacturing & Service Operations Management*. 15(4): 689–700.

[164] Drews, J. (2000). "Drug discovery: a historical perspective". *Science*. 287(5460): 1960–1964.

[165] Du Boff, R. B. (1967). "The introduction of electric power in American manufacturing". *The Economic History Review*. 20(3): 509–518.

[166] Duhan, J. S., R. Kumar, N. Kumar, P. Kaur, K. Nehra, and S. Duhan. (2017). "Nanotechnology: The new perspective in precision agriculture". *Biotechnology Reports*. 15: 11–23.

[167] Duistermaat, K. (2008). "The pots and potters of Assyria". *Technology and Organisation of Production, Ceramic Sequence and Vessel Function at Late Bronze Age Tell Sabi Abyad, Syria.*

[168] Dunham, D. (1956). *Building an Egyptian pyramid.* Verlag nicht ermittelbar.

[169] Economic – Department of Economic and Social Affairs, United Nations. "The 17 Goals". URL: https://sdgs.un.org/goals (accessed on 05/27/2023).

[170] Economic – Bureau of Economic Analysis, United States. (2020). "Gross Domestic Product by Industry Fourth Quarter and Year 2019". URL: https://www.bea.gov/system/files/2020-04/gdpind419.pdf.

[171] Edwards, A. R. (2005). *The sustainability revolution: Portrait of a paradigm shift*. New Society Publishers.

[172] Edwards, J. F. (2003). "Building the great pyramid: Probable construction methods employed at Giza". *Technology and Culture*. 44(2): 340–354.

[173] Efficiency, E. (2007). "Tracking industrial energy efficiency and CO2 emissions". *International Energy Agency*. 34(2): 1–12.

[174] EIA – US Energy Information Administration. (2011). "Annual energy review". *Energy Information Administration, US Department of Energy: Washington, DC www.eia.doe.gov/emeu/aer*: 80.

[175] Eichengreen, B. and P. Gupta. (2013a). "Exports of services: Indian experience in perspective". *Indian Growth and Development Review*.

[176] Eichengreen, B. and P. Gupta. (2013b). "The two waves of service-sector growth". *Oxford Economic Papers*. 65(1): 96–123.

[177] Eisenstein, E. L., E. Elizabeth Lewisohn, *et al.* (2005). *The printing revolution in early modern Europe*. Cambridge University Press.

[178] Elkington, J. (1998). "Partnerships from cannibals with forks: The triple bottom line of 21st-century business". *Environmental quality management*. 8(1): 37–51.

[179] Englund, R. K. (2004). "Proto-cuneiform account-books and journals". *Creating Economic Order*: 23–46.

[180] Environment, World Commission on Environment. (1992). *Our common future*. Centre for Our Common Future.

[181] Erb-Satullo, N. L. (2019). "The innovation and adoption of iron in the ancient Near East". *Journal of Archaeological Research*. 27(4): 557–607.

[182] Erdoğdu, M. M. (2023). "Lights-Out Manufacturing and Foreign Direct Investment Decline: Human Resource-Based Avenues for Technological Diffusion in Developing Countries". In: *The Political Economy of Global Manufacturing, Business and Finance*. Springer. 97–119.

[183] Escoffier, M. R. (1987). "The chef in society: Origins and development". *Hospitality Review*. 5(1): 6.

[184] Eurostat. "Gross value added in the EU by economic activity". URL: https://ec.europa.eu/eurostat/statistics-explained/index.php?title=National_accounts_and_GDP#Gross_value_added_in_the_EU_by_economic_activity (accessed on 01/04/2023).

[185] Evans, A. M. (2009). *An introduction to economic geology and its environmental impact*. John Wiley & Sons.

[186] Ewijk, S. van, J. A. Stegemann, and P. Ekins. (2021). "Limited climate benefits of global recycling of pulp and paper". *Nature Sustainability*. 4(2): 180–187.

[187] Fabregas, R., M. Kremer, and F. Schilbach. (2019). "Realizing the potential of digital development: The case of agricultural advice". *Science*. 366(6471): eaay3038.

[188] Fang, R., Y.-F. Yu, E.-J. Li, N.-X. Lv, Z.-C. Liu, H.-G. Zhou, and X.-D. Song. (2022). "Global, regional, national burden and gender disparity of cataract: findings from the global burden of disease study 2019". *BMC Public Health*. 22(1): 2068.

[189] Fara, P. (1999). "Catch a falling apple: Isaac Newton and myths of genius". *Endeavour*. 23(4): 167–170.

[190] Farley, T. (2005). "Mobile telephone history". *Telektronikk*. 101(3/4): 22.

[191] Farooque, M., A. Zhang, M. Thürer, T. Qu, and D. Huisingh. (2019). "Circular supply chain management: A definition and structured literature review". *Journal of cleaner production*. 228: 882–900.

[192] Farrell, J., P. B. Burow, K. McConnell, J. Bayham, K. Whyte, and G. Koss. (2021). "Effects of land dispossession and forced migration on Indigenous peoples in North America". *Science*. 374(6567): eabe4943.

[193] Febvre, L. and H.-J. Martin. (1997). *The coming of the book: the impact of printing 1450–1800*: Vol. 10. Verso.

[194] Feeney, J. (2019). "Hunter-gatherer land management in the human break from ecological sustainability". *The Anthropocene Review*. 6(3): 223–242.

[195] Feng, L., Y. Hu, C. A. Hall, and J. Wang. (2012). *The Chinese Oil Industry: History and Future*. Springer Science & Business Media.

[196] Ferguson, R. B. (2013). "The prehistory of war and peace in Europe and the Near East". *War, peace, and human nature: the convergence of evolutionary and cultural views*: 191–240.

[197] Fineberg, H. V. (2012). "A successful and sustainable health system - how to get there from here". *New England Journal of Medicine*. 366(11): 1020–1027.

[198] Fischedick, M., J. Roy, A. Abdel-Aziz, A. A., and *et al.* (2014). "Climate Change 2014: Mitigation of Climate Change." In: ed. by O. Edenhofer, Pichs-Madruga, and *et al.* Cambridge University Press, Cambridge, United Kingdom and New York, NY, USA. Chap. Industry. Contribution of Working Group III to the Fifth Assessment Report of the Intergovernmental Panel on Climate Change.

[199] Fisher, A. G. (1933). "Capital and the Growth of Knowledge". *The Economic Journal*: 379–389.

[200] Flinn, M. W. (1959). "Abraham Darby and the coke-smelting process". *Economica*. 26(101): 54–59.

[201] Florek, S. and R. Young. (2020). "Confederate Calamity: The Rise and Use of the Telegraph During the Civil War".

[202] Ford, H. (2019). *Today and Tomorrow: Commemorative Edition of Ford's 1926 Classic*. Routledge.

[203] Ford. The Henry Ford Website. URL: https://www.thehenryford.org/explore/stories-of-innovation/visionaries/henry-ford (accessed on 08/26/2021).

[204] Føre, M., K. Frank, T. Norton, E. Svendsen, J. A. Alfredsen, T. Dempster, H. Eguiraun, W. Watson, A. Stahl, L. M. Sunde, *et al.* (2018). "Precision fish farming: A new framework to improve production in aquaculture". *biosystems engineering*. 173: 176–193.

[205] Fraiman, N. M., L. V. Green, A. Heching, and G. J. Van Ryzin. (2010). "The Ritz-Carlton Hotel Company The Quest for Service Excellence".

[206] Frangipane, M. (2007). "Different types of egalitarian societies and the development of inequality in early Mesopotamia". *World Archaeology*. 39(2): 151–176.

[207] Freeman, R. B. and R. Schettkat. (2005). "Marketization of household production and the EU–US gap in work". *Economic policy*. 20(41): 6–50.

[208] Frey, C. B. and M. Osborne. (2013). "The future of employment".

[209] Fukuyama, M. (2018). "Society 5.0: Aiming for a new human-centered society". *Japan Spotlight*. 27(5): 47–50.

[210] Fund, The King's Fund. "Key Facts and Figures". URL: https://www.kingsfund.org.uk/audio-video/key-facts-figures-nhs (accessed on 02/12/2023).

[211] Gabrial, B. (2008). "History of writing technologies". *Handbook of research on writing: history, society, school, individual, text*: 27–40.

[212] Gadgil, M. (1990). "India's deforestation: patterns and processes". *Society & Natural Resources*. 3(2): 131–143.

[213] Gaj, K. and A. Orłowski. (2003). "Facts and myths of enigma: Breaking stereotypes". In: *International Conference on the Theory and Applications of Cryptographic Techniques*. Springer. 106–122.

[214] Gallai, N., J.-M. Salles, J. Settele, and B. E. Vaissière. (2009). "Economic valuation of the vulnerability of world agriculture confronted with pollinator decline". *Ecological economics*. 68(3): 810–821.

[215] Garza-Reyes, J. A. (2015). "Lean and green–a systematic review of the state of the art literature". *Journal of cleaner production*. 102: 18–29.

[216] Gass, S. I. and A. A. Assad. (2005). *An annotated timeline of operations research: An informal history*. Vol. 75. Springer Science & Business Media.

[217] Gavronski, I., E. L. Paiva, R. Teixeira, and M. C. F. de Andrade. (2013). "ISO 14001 certified plants in Brazil–taxonomy and practices". *Journal of Cleaner Production*. 39: 32–41.

[218] Georgiou, L., C. J. Dunmore, A. Bardo, L. T. Buck, J.-J. Hublin, D. H. Pahr, D. Stratford, A. Synek, T. L. Kivell, and M. M. Skinner. (2020). "Evidence for habitual climbing in a Pleistocene hominin in South Africa". *Proceedings of the National Academy of Sciences*. 117(15): 8416–8423.

[219] Gereffi, G., A. C. Posthuma, and A. Rossi. (2021). "Introduction: Disruptions in global value chains–Continuity or change for labour governance?"

[220] German, L. A., A. M. Bonanno, L. C. Foster, and L. Cotula. (2020). ""Inclusive business" in agriculture: Evidence from the evolution of agricultural value chains". *World Development*. 134: 105018.

[221] Gilberth, F. B. and L. Gilbreth. (1921). "Process Chart".

[222] Gilbreth, F. B. and L. M. Gilbreth. (1924). "Classifying the elements of work". *Management and Administration*. 8(2): 151–154.

[223] Gilbreth, L. M. *et al.* (1913). "The psychology of management". *Industrial Engineering*.

[224] Gill, J. L. (2014). "Ecological impacts of the l ate Q uaternary megaherbivore extinctions". *New Phytologist*. 201(4): 1163–1169.

[225] Gimenez, C., V. Sierra, and J. Rodon. (2012). "Sustainable operations: Their impact on the triple bottom line". *International Journal of Production Economics*. 140(1): 149–159.

[226] Girotto, F., L. Alibardi, and R. Cossu. (2015). "Food waste generation and industrial uses: A review". *Waste management*. 45: 32–41.

[227] Glazko, G. V. and M. Nei. (2003). "Estimation of Divergence Times for Major Lineages of Primate Species". *Molecular Biology and Evolution*. 20(3): 424–434.

[228] Glouberman, S. (2009). "Knowledge transfer and the complex story of scurvy". *Journal of evaluation in clinical practice*. 15(3): 553–557.

[229] Gnesin, G. (2015). "Metals and Alloys of Bronze Age: from Middle to Modern Times. II. Gold, Silver, Tin, Lead, Mercury, and Their Alloys". *Powder Metallurgy and Metal Ceramics*. 53(11–12): 722–732.

[230] Gochfeld, M. (2005). "Occupational medicine practice in the United States since the industrial revolution". *Journal of occupational and environmental medicine*. 47(2): 115–131.

[231] Goldfarb, B. (2018). *Eager: The surprising, secret life of beavers and why they matter*. Chelsea Green Publishing.

[232] Goldin, C. and R. A. Margo. (1992). "The great compression: The wage structure in the United States at mid-century". *The Quarterly Journal of Economics*. 107(1): 1–34.

[233] Goldstine, H. H. and A. Goldstine. (1946). "The electronic numerical integrator and computer (eniac)". *Mathematical Tables and Other Aids to Computation*. 2(15): 97–110.

[234] Goldstone, J. A. (2002). "Efflorescences and economic growth in world history: rethinking the "Rise of the West" and the Industrial Revolution". *Journal of world history*: 323–389.

[235] Golonka, J. and A. Gaweda. (2012). "Plate Tectonic Evolution of the Southern Margin of Laurussia in the Paleozoic". *Intech Tectonics-recent advances*. 10: 262–282.

[236] Gonzalez, C. G. (2002). "Seasons of resistance: sustainable agriculture and food security in Cuba". *Tul. Envtl. LJ*. 16: 685.

[237] Goulson, D., G. C. Lye, and B. Darvill. (2008). "Decline and conservation of bumble bees". *Annu. Rev. Entomol.* 53: 191–208.

[238] Graham, L. D. (1999). "Domesticating efficiency: Lillian Gilbreth's scientific management of homemakers, 1924–1930". *Signs: Journal of Women in Culture and Society*. 24(3): 633–675.

[239] Grainger, C. and K. Beauchemin. (2011). "Can enteric methane emissions from ruminants be lowered without lowering their production?" *Animal feed science and technology*. 166: 308–320.

[240] Grattan, J., S. Huxley, L. A. Karaki, H. Toland, D. Gilbertson, B. Pyatt, and Z. Al Saad. (2002). "'Death... more desirable than life'? The human skeletal record and toxicological implications of ancient copper mining and smelting in Wadi Faynan, southwestern Jordan". *Toxicology and industrial health*. 18(6): 297–307.

[241] Gregory, C. E. (2021). *A concise history of mining*. CRC Press.

[242] Grigg, D. B. (1975). "The World's Agricultural Labour Force 1800–1970". *Geography*: 194–202.

[243] Griliches, Z. (2008). *Output measurement in the service sectors*. Vol. 56. University of Chicago Press.

[244] Groucutt, H. S., M. D. Petraglia, G. Bailey, E. M. Scerri, A. Parton, L. Clark-Balzan, R. P. Jennings, L. Lewis, J. Blinkhorn, N. A. Drake, *et al.* (2015). "Rethinking the dispersal of Homo sapiens out of Africa". *Evolutionary Anthropology: Issues, News, and Reviews*. 24(4): 149–164.

[245] Grubb, M., N. D. Jordan, E. Hertwich, K. Neuhoff, K. Das, K. R. Bandyopadhyay, H. Van Asselt, M. Sato, R. Wang, W. A. Pizer, *et al.* (2022). "Carbon leakage, consumption, and trade". *Annual Review of Environment and Resources*. 47: 753–795.

[246] Gurven, M. and K. Hill. (2009). "Why do men hunt? A reevaluation of "man the hunter" and the sexual division of labor". *Current Anthropology*. 50(1): 51–74.

[247] Gustavsson, J., C. Cederberg, U. Sonesson, R. Van Otterdijk, and A. Meybeck. (2011). "Global food losses and food waste–Extent, causes and prevention".

[248] Haidari, L. A., S. T. Brown, M. Ferguson, E. Bancroft, M. Spiker, A. Wilcox, R. Ambikapathi, V. Sampath, D. L. Connor, and B. Y. Lee. (2016). "The economic and operational value of using drones to transport vaccines". *Vaccine*. 34(34): 4062–4067.

[249] Haidle, M. N. (2010). "Working-memory capacity and the evolution of modern cognitive potential: implications from animal and early human tool use". *Current Anthropology*. 51(S1): S149–S166.

[250] Hák, T., S. Janoušková, and B. Moldan. (2016). "Sustainable Development Goals: A need for relevant indicators". *Ecological indicators*. 60: 565–573.

[251] Hale, R. C., M. E. Seeley, M. J. La Guardia, L. Mai, and E. Y. Zeng. (2020). "A global perspective on microplastics". *Journal of Geophysical Research: Oceans*. 125(1): e2018JC014719.

[252] Hammond, M. B., P. Miller, E. Whitney, and N. Pendleton. (1897). "Correspondence of Eli Whitney relative to the Invention of the Cotton Gin". *The American Historical Review*. 3(1): 90–127.

[253] Han, J.-W. and M. Meyyappan. (2014). "The device made of nothing". *IEEE Spectrum*. 51(7): 30–35.

[254] Handfield, R. B., M. Onitsuka, *et al.* (1995). "Process and Supply Chain Evolution in the American Cotton Textile Industry". *St. Andrew's University Economic and Business Review*. 37(3): 1–35.

[255] Hanlon, W. W. (2020). "Coal smoke, city growth, and the costs of the industrial revolution". *The Economic Journal*. 130(626): 462–488.

[256] Harnett, B. (2017). "The diffusion of the codex". *Classical Antiquity*. 36(2): 183–235.

[257] Hassan, A. (2013). "Review of the global oil and gas industry: a concise journey from ancient time to modern world". *Petroleum Technology Development Journal*. 3(2): 123–141.

[258] Headland, T. N., L. A. Reid, M. Bicchieri, C. A. Bishop, R. Blust, N. E. Flanders, P. M. Gardner, K. L. Hutterer, A. Marciniak, R. F. Schroeder, *et al.* (1989). "Hunter-gatherers and their neighbors from prehistory to the present [and comments and replies]". *Current anthropology*. 30(1): 43–66.

[259] Heitmann, J. A. (1987). "The modernization of the Louisiana sugar industry, 1830–1910".

[260] Heizer, J., B. Render, and C. Munson. (2017). *Principles of Operations Management: Sustainability and Supply Chain Management (Global editon)*. Pearson.

[261] Henn, J. J., C. B. Anderson, and G. Martínez Pastur. (2016). "Landscape-level impact and habitat factors associated with invasive beaver distribution in Tierra del Fuego". *Biological Invasions*. 18: 1679–1688.

[262] Herrendorf, B., R. Rogerson, and A. Valentinyi. (2014). "Growth and structural transformation". *Handbook of economic growth*. 2: 855–941.

[263] Hicks, D. A. (1997). "The inequality-adjusted human development index: a constructive proposal". *World development*. 25(8): 1283–1298.

[264] Higgins-Desbiolles, F. (2018). "Sustainable tourism: Sustaining tourism or something more?" *Tourism management perspectives*. 25: 157–160.

[265] Hill, D. R. (1991). "Mechanical engineering in the medieval near east". *Scientific American*. 264(5): 100–105.

[266] Hill, P. (2012). *The Book of Knowledge of Ingenious Mechanical Devices: Kitab fima rifat al-hiyal al-handasiyya*. Springer Science & Business Media.

[267] Hobbs, F. (2002). *Demographic trends in the 20th century*. Vol. 4. US Department of Commerce, Economics and Statistics Administration, US ....

[268] Hoffecker, J. F. (2009). "The spread of modern humans in Europe". *Proceedings of the National Academy of Sciences*. 106(38): 16040–16045.

[269] Hoffmann, S., I. Balderjahn, B. Seegebarth, R. Mai, and M. Peyer. (2018). "Under which conditions are consumers ready to boycott or buycott? The roles of hedonism and simplicity". *Ecological Economics*. 147: 167–178.

[270] Holden, C. *et al.* (2004). "The origin of speech". *Science*. 303(5662): 1316–1319.

[271] Hollerith, H. (1894). "The electrical tabulating machine". *Journal of the Royal Statistical Society*. 57(4): 678–689.

[272] Holweg, M. (2007). "The genealogy of lean production". *Journal of operations management*. 25(2): 420–437.

[273] Hospitals – Aravind Eye Care System. "Detailed Performance Report 2021–2022". URL: https://aravind.org/wp-content/uploads/2022/09/Performance-2021-2022.pdf (accessed on 02/12/2023).

[274] Hotels – Ritz Carlton. "Gold Standards". URL: https://www.ritzcarlton.com/en/about/gold-standards (accessed on 02/09/2023).

[275] Houston, R. "Model T Ford Production". URL: https://www.mtfca.com/encyclo/fdprod.htm (accessed on 02/09/2021).

[276] Howard, R. A. (1978). "Interchangeable parts reexamined: The private sector of the American arms industry on the eve of the civil war". *Technology and Culture*: 633–649.

[277] Hsieh, C.-T. and E. Rossi-Hansberg. (2023). "The industrial revolution in services". *Journal of Political Economy Macroeconomics*. 1(1): 3–42.

[278] Humphries, J. and J. Weisdorf. (2019). "Unreal wages? Real income and economic growth in England, 1260–1850". *The Economic Journal*. 129(623): 2867–2887.

[279] Hurley, A. (1994). "Creating Ecological Wastelands: Oil Pollution in New York City, 1870–1900". *Journal of Urban History*. 20(3): 340–364.

[280] Huta, V. (2016). "An overview of hedonic and eudaimonic well-being concepts". *The Routledge handbook of media use and well-being*: 14–33.

[281] IBM. "IBM Archives, Workhorse of modern industry: The IBM 650". URL: https://www.ibm.com/ibm/history/exhibits/650/650_intro.html (accessed on 10/20/2021).

[282] IRENA, International Renewable Energy Agency. (2022). *World energy transitions outlook*. eBook Partnership.

[283] Israel, P. B. (2002). "Inventing industrial research: Thomas Edison and the Menlo Park laboratory". *Endeavour*. 26(2): 48–54.

[284] Jackson, D. E. and F. L. Ratnieks. (2006). "Communication in ants". *Current biology*. 16(15): R570–R574.

[285] Jacobs, F. R. *et al.* (2007). "Enterprise resource planning (ERP)—A brief history". *Journal of operations management*. 25(2): 357–363.

[286] Jaehn, F. (2016). "Sustainable operations". *European Journal of Operational Research*. 253(2): 243–264.

[287] John, A., C. Clark, M. Freeman, K. Kerrisk, S. Garcia, and I. Halachmi. (2016). "Milking robot utilization, a successful precision livestock farming evolution". *Animal*. 10(9): 1484–1492.

[288] Johnson, C. N. (2009). "Ecological consequences of Late Quaternary extinctions of megafauna". *Proceedings of the Royal Society B: Biological Sciences*. 276(1667): 2509–2519.

[289] Jones, C. G., J. H. Lawton, and M. Shachak. (1997). "Positive and negative effects of organisms as physical ecosystem engineers". *Ecology*. 78(7): 1946–1957.

[290] Jordan, P., K. Gibbs, P. Hommel, H. Piezonka, F. Silva, and J. Steele. (2016). "Modelling the diffusion of technologies across Afro-Eurasia: emerging insights and future research". *Antiquity*. 90(351): 590–603.

[291] Kalimeris, P., K. Bithas, C. Richardson, and P. Nijkamp. (2020). "Hidden linkages between resources and economy: A "Beyond-GDP" approach using alternative welfare indicators". *Ecological Economics*. 169: 106508.

[292] Kampa, M. and E. Castanas. (2008). "Human health effects of air pollution". *Environmental pollution*. 151(2): 362–367.

[293] Katouzian, M. (1970). "The development of the service sector: a new approach". *Oxford Economic Papers*. 22(3): 362–382.

[294] Kennedy, A. B. and H. R. Sankey. (1898). "The Thermal Efficiency of Steam Engines". In: *Minutes of the Proceedings of the Institution of Civil Engineers*, Vol. 134. pp. 278–312.

[295] Khan, I., K. Saeed, and I. Khan. (2019). "Nanoparticles: Properties, applications and toxicities". *Arabian journal of chemistry*. 12(7): 908–931.

[296] Khan, S. A., M. Koç, and S. G. Al-Ghamdi. (2021). "Sustainability assessment, potentials and challenges of 3D printed concrete structures: A systematic review for built environmental applications". *Journal of Cleaner Production*. 303: 127027.

[297] Khan, Z. and K. L. Sokoloff. (2004). "Institutions and democratic invention in 19th-century america: Evidence from "great inventors," 1790–1930". *American Economic Review*. 94(2): 395–401.

[298] King, L. (2011). "The Avalon Project". *Documents in law, history and diplomacy. Code of Hammurabi*. Online: http://avalon.law.yale.edu/ancient/hamframe.asp. Accessed. 26.

[299] Kirchherr, J., D. Reike, and M. Hekkert. (2017). "Conceptualizing the circular economy: An analysis of 114 definitions". *Resources, conservation and recycling*. 127: 221–232.

[300] Kittler, F. A. (1999). *Gramophone, film, typewriter*. Stanford University Press.

[301] Klein Goldewijk, K., A. Beusen, J. Doelman, and E. Stehfest. (2017). "Anthropogenic land use estimates for the Holocene–HYDE 3.2". *Earth System Science Data*. 9(2): 927–953.

[302] Klein Goldewijk, K., A. Beusen, and P. Janssen. (2010). "Long-term dynamic modeling of global population and built-up area in a spatially explicit way: HYDE 3.1". *The Holocene*. 20(4): 565–573.

[303] Kleindorfer, P. R., K. Singhal, and L. N. Van Wassenhove. (2005). "Sustainable operations management". *Production and operations management*. 14(4): 482–492.

[304] KNOEMA. "Top Vehicle Manufacturers in the US Market, 1961–2016". URL: https://knoema.com/infographics/floslle/top-vehicle-manufacturers-in-the-us-market-1961-2016 (accessed on 02/08/2023).

[305] Koberg, E. and A. Longoni. (2019). "A systematic review of sustainable supply chain management in global supply chains". *Journal of cleaner production*. 207: 1084–1098.

[306] Koch, P. L. and A. D. Barnosky. (2006). "Late Quaternary extinctions: state of the debate". *Annual Review of Ecology, Evolution, and Systematics*. 37.

[307] Koh, L., G. Orzes, and F. J. Jia. (2019). "The fourth industrial revolution (Industry 4.0): technologies disruption on operations and supply chain management". *International Journal of Operations & Production Management*.

[308] Kolk, A. and R. Van Tulder. (2002). "Child labor and multinational conduct: a comparison of international business andstakeholder codes". *Journal of Business Ethics*. 36: 291–301.

[309] Konings, J. and A. P. Murphy. (2006). "Do multinational enterprises relocate employment to low-wage regions? Evidence from European multinationals". *Review of World Economics*. 142: 267–286.

[310] Kooij, B. J. van der. (2021). "How did the General Purpose Technology 'Internal Combustion Engine' contribute to the Third Industrial Revolution: The Power Engines." *Available at SSRN 3852628*.

[311] Kowalkowski, C., H. Gebauer, B. Kamp, and G. Parry. (2017). "Servitization and deservitization: Overview, concepts, and definitions". *Industrial Marketing Management*. 60: 4–10.

[312] Krajewski, L. J., L. P. Ritzman, and M. K. Malhotra. (2010). *Operations management: Processes and supply chains*. Pearson Upper Saddle River, New Jersey.

[313] Kriegler, E., J. P. Weyant, G. J. Blanford, V. Krey, L. Clarke, J. Edmonds, A. Fawcett, G. Luderer, K. Riahi, R. Richels, *et al.* (2014). "The role of technology for achieving climate policy objectives: overview of the EMF 27 study on global technology and climate policy strategies". *Climatic Change*. 123: 353–367.

[314] Krishna, K., B. Goldar, D. K. Das, S. C. Aggarwal, A. A. Erumban, and P. C. Das. (2022). "India Productivity Report".

[315] Kritsky, G. (2017). "Beekeeping from antiquity through the middle ages". *Annual Review of Entomology*. 62: 249–264.

[316] Kubiszewski, I., R. Costanza, C. Franco, P. Lawn, J. Talberth, T. Jackson, and C. Aylmer. (2013). "Beyond GDP: Measuring and achieving global genuine progress". *Ecological economics*. 93: 57–68.

[317] Kuhn, D. (1979). "The Spindle-Wheel: A Chou Chinese Invention". *Early China*. 5: 14–24.

[318] Kuhn, S. and M. Stiner. (2006). "What'sa mother to do? The division of labor among Neandertals and modern humans in Eurasia". *Current anthropology*. 47(6): 953–981.

[319] Kumar, P., R. K. Singh, and V. Kumar. (2021). "Managing supply chains for sustainable operations in the era of industry 4.0 and circular economy: Analysis of barriers". *Resources, Conservation and Recycling*. 164: 105215.

[320] Labor Statistics – Bureau for Labor Statistics. "Emloyment by Major Industry". URL: https://www.bls.gov/emp/tables/employment-by-major-industry-sector.htm (accessed on 01/05/2023).

[321] Laboratory – The Global Monitoring Laboratory of the National Oceanic and Atmospheric Administration. "Trends in Atmospheric Carbon Dioxide". URL: https://gml.noaa.gov/ccgg/trends/global.html (accessed on 05/16/2023).

[322] Lade, S. J., W. Steffen, W. De Vries, S. R. Carpenter, J. F. Donges, D. Gerten, H. Hoff, T. Newbold, K. Richardson, and J. Rockström. (2020). "Human impacts on planetary boundaries amplified by Earth system interactions". *Nature sustainability*. 3(2): 119–128.

[323] Lakner, C., D. G. Mahler, M. Negre, and E. B. Prydz. (2022). "How much does reducing inequality matter for global poverty?" *The Journal of Economic Inequality*. 20(3): 559–585.

[324] Lakwete, A. (2005). *Inventing the cotton gin: machine and myth in Antebellum America*. JHU Press.

[325] Laland, K., B. Matthews, and M. W. Feldman. (2016). "An introduction to niche construction theory". *Evolutionary ecology*. 30(2): 191–202.

[326] Laland, K. N. (2004). "Extending the extended phenotype". *Biology and Philosophy*. 19(3): 313–325.

[327] Lamb, W. F., T. Wiedmann, J. Pongratz, R. Andrew, M. Crippa, J. G. Olivier, D. Wiedenhofer, G. Mattioli, A. Al Khourdajie, J. House, *et al.* (2021). "A review of trends and drivers of greenhouse gas emissions by sector from 1990 to 2018". *Environmental research letters*. 16(7): 073005.

[328] Landrigan, P. J., R. Fuller, N. J. Acosta, O. Adeyi, R. Arnold, A. B. Baldé, R. Bertollini, S. Bose-O'Reilly, J. I. Boufford, P. N. Breysse, *et al.* (2018). "The Lancet Commission on pollution and health". *The lancet*. 391(10119): 462–512.

[329] Lasi, H., P. Fettke, H.-G. Kemper, T. Feld, and M. Hoffmann. (2014). "Industry 4.0". *Business & information systems engineering*. 6: 239–242.

[330] Lassaletta, L., G. Billen, J. Garnier, L. Bouwman, E. Velazquez, N. D. Mueller, and J. S. Gerber. (2016). "Nitrogen use in the global food system: past trends and future trajectories of agronomic performance, pollution, trade, and dietary demand". *Environmental Research Letters.* 11(9): 095007.

[331] Leach, M., K. Raworth, and J. Rockström. (2013). "Between social and planetary boundaries: Navigating pathways in the safe and just space for humanity".

[332] Leakey, M. D. (1971). *Olduvai Gorge: Volume 3, excavations in beds I and II, 1960–1963*, Vol. 3. Cambridge University Press.

[333] Lee, F. (2004). *If Disney ran your hospital: 9 1/2 things you would do differently.* Second River Healthcare Press Bozeman, MT.

[334] Lee, R. B. (1979). *The! Kung San: men, women and work in a foraging society.* CUP Archive.

[335] Legault, L. (2020). "Intrinsic and extrinsic motivation". *Encyclopedia of personality and individual differences*: 2416–2419.

[336] Leiner, B. M., V. G. Cerf, D. D. Clark, R. E. Kahn, L. Kleinrock, D. C. Lynch, J. Postel, L. G. Roberts, and S. S. Wolff. (1997). "The past and future history of the Internet". *Communications of the ACM.* 40(2): 102–108.

[337] Lenzen, M., A. Malik, M. Li, J. Fry, H. Weisz, P.-P. Pichler, L. S. M. Chaves, A. Capon, and D. Pencheon. (2020). "The environmental footprint of health care: a global assessment". *The Lancet Planetary Health.* 4(7): e271–e279.

[338] Lenzen, M., Y.-Y. Sun, F. Faturay, Y.-P. Ting, A. Geschke, and A. Malik. (2018). "The carbon footprint of global tourism". *Nature climate change.* 8(6): 522–528.

[339] Leroi-Gourhan, A. (1982). "The archaeology of Lascaux cave". *Scientific American.* 246(6): 104–113.

[340] Levchenko, I., S. Xu, S. Mazouffre, M. Keidar, and K. Bazaka. (2021). "Mars colonization: beyond getting there". *Terraforming Mars*: 73–98.

[341] Levy, J. S. and W. R. Thompson. (2011). *The arc of war: Origins, escalation, and transformation.* University of Chicago Press.

[342] Lewis, H. M., L. Vinicius, J. Strods, R. Mace, and A. B. Migliano. (2014). "High mobility explains demand sharing and enforced cooperation in egalitarian hunter-gatherers". *Nature Communications.* 5(1): 1–8.

[343] Lindert, P. H. and J. G. Williamson. (1985). "Growth, equality, and history". *Explorations in Economic History.* 22(4): 341–377.

[344] Lindert, P. H. and J. G. Williamson. (2016). "Unequal gains: American growth and inequality since 1700". *Juncture.* 22: 276–283.

[345] Lipinski, B., C. O'Connor, and C. Hanson. (2016). "SDG target 12.3 on food loss and waste: 2016 progress report". *Champions, Washington, DC.*

[346] Liu, Y., X. Ma, L. Shu, G. P. Hancke, and A. M. Abu-Mahfouz. (2020). "From Industry 4.0 to Agriculture 4.0: Current status, enabling technologies, and research challenges". *IEEE Transactions on Industrial Informatics*. 17(6): 4322–4334.

[347] Liu, Z., Z. Deng, S. Davis, and P. Ciais. (2023). "Monitoring global carbon emissions in 2022". *Nature Reviews Earth & Environment*. 4(4): 205–206.

[348] Lloyd, C. C. (1963). "The conquest of scurvy". *The British Journal for the History of Science*. 1(4): 357–363.

[349] Lombard, M. (2016). "Mountaineering or ratcheting? Stone Age hunting weapons as proxy for the evolution of human technological, behavioral and cognitive flexibility". In: *The naturMe of culture*. Springer. 135–146.

[350] Longoni, A. and R. Cagliano. (2015). "Environmental and social sustainability priorities: Their integration in operations strategies". *International Journal of Operations & Production Management*. 35(2): 216–245.

[351] Lowry, G. V., A. Avellan, and L. M. Gilbertson. (2019). "Opportunities and challenges for nanotechnology in the agri-tech revolution". *Nature nanotechnology*. 14(6): 517–522.

[352] Lu, B., J. Zhang, D. Xue, and P. Gu. (2011). "Systematic lifecycle design for sustainable product development". *Concurrent Engineering*. 19(4): 307–324.

[353] Lu, V. N., J. Wirtz, W. H. Kunz, S. Paluch, T. Gruber, A. Martins, and P. G. Patterson. (2020). "Service robots, customers and service employees: what can we learn from the academic literature and where are the gaps?" *Journal of Service Theory and Practice*. 30(3): 361–391.

[354] Lukasik, S. (2010). "Why the ARPANET was built". *IEEE Annals of the History of Computing*. 33(3): 4–21.

[355] Lund-Thomsen, P. and A. Lindgreen. (2014). "Corporate social responsibility in global value chains: Where are we now and where are we going?" *Journal of Business Ethics*. 123: 11–22.

[356] MacDougall, A. H. and P. Friedlingstein. (2015). "The origin and limits of the near proportionality between climate warming and cumulative CO2 emissions". *Journal of Climate*. 28(10): 4217–4230.

[357] Maddison, A. *et al.* (2008). "The west and the rest in the world economy: 1000–2030". *World Economics*. 9(4): 75–99.

[358] Madsen, J. B. and F. Murtin. (2017). "British economic growth since 1270: the role of education". *Journal of Economic Growth*. 22(3): 229–272.

[359] Madsen, J. B. *et al.* (2016). "Human accomplishment and growth in Britain since 1270: The role of great scientists and education". *Tech. rep.* Monash University, Department of Economics.

[360] Magkos, F., I. Tetens, S. G. Bügel, C. Felby, S. R. Schacht, J. O. Hill, E. Ravussin, and A. Astrup. (2020). "The environmental foodprint of obesity". *Obesity*. 28(1): 73–79.

[361] Mahajan, R. P. (2011). "The WHO surgical checklist". *Best Practice & Research Clinical Anaesthesiology*. 25(2): 161–168.

[362] Malmodin, J. and D. Lundén. (2018). "The energy and carbon footprint of the global ICT and E&M sectors 2010–2015". *Sustainability*. 10(9): 3027.

[363] Marlowe, F. W., J. C. Berbesque, B. Wood, A. Crittenden, C. Porter, and A. Mabulla. (2014). "Honey, Hadza, hunter-gatherers, and human evolution". *Journal of human evolution*. 71: 119–128.

[364] Martin, C., D. P. Morgavi, and M. Doreau. (2010). "Methane mitigation in ruminants: from microbe to the farm scale". *Animal*. 4(3): 351–365.

[365] Maslow, A. and K. Lewis. (1987). "Maslow's hierarchy of needs". *Salenger Incorporated*. 14(17): 987–990.

[366] Maw, P., P. Solar, A. Kane, and J. S. Lyons. (2021). "After the great inventions: technological change in UK cotton spinning, 1780–1835". *The Economic History Review*.

[367] Mayewski, P. A., E. E. Rohling, J. C. Stager, W. Karlén, K. A. Maasch, L. D. Meeker, E. A. Meyerson, F. Gasse, S. van Kreveld, K. Holmgren, *et al.* (2004). "Holocene climate variability". *Quaternary research*. 62(3): 243–255.

[368] Mayor, S. (2016). "Sixty seconds on... scurvy".

[369] Mazwane, S., M. N. Makhura, and M. P. Senyolo. (2022). "Important Policy Parameters for the Development of Inclusive Digital Agriculture: Implications for the Redistributive Land Reform Program in South Africa". *Agriculture*. 12(12): 2129.

[370] McCall, B. (2019). "Sub-Saharan Africa leads the way in medical drones". *The Lancet*. 393(10166): 17–18.

[371] McClelland, P. D. (1997). *Sowing modernity: America". first agricultural revolution*. Cornell University Press.

[372] McCown, E. R. and D. A. Hamburg. (1979). *The Great Apes*. Benjamin/Cummings.

[373] McCusker, J. J. (2000). "Estimating early American gross domestic product". *Historical Methods: A Journal of Quantitative and Interdisciplinary History*. 33(3): 155–162.

[374] McElwain, J. C. and S. W. Punyasena. (2007). "Mass extinction events and the plant fossil record". *Trends in ecology & evolution*. 22(10): 548–557.

[375] Mckinnon, A. C. (2016). "The possible impact of 3D printing and drones on last-mile logistics: An exploratory study". *Built Environment*. 42(4): 617–629.

[376] McNeill, J. R. (2004). "Woods and warfare in world history". *Environmental History*. 9(3): 388–410.

[377] McNurlin, B. C. and R. H. Sprague. (2005). *Information systems management in practice*. Prentice-Hall, Inc.

[378] Meert, J. G. and R. Van Der Voo. (1997). "The assembly of Gondwana 800-550 Ma". *Journal of Geodynamics*. 23(3–4): 223–235.

[379] Mehta Pavitra, K. and S. Shenoy. (2011). *Infinite Vision*. Berret-Koehler Publishers.

[380] Merriam-Webster. "Dictionary". URL: https://www.merriam-webster.com (accessed on 06/14/2021).

[381] Michener, C. D. (2000). *The bees of the world*. Vol. 1. JHU press.

[382] Milanovic, B. (2022). "The three eras of global inequality, 1820–2020 with the focus on the past thirty years".

[383] Mills, C. (2016). *Regulating health and safety in the British mining industries, 1800–1914*. Routledge.

[384] Mirzabaev, A., J. Wu, J. Evans, F. García-Oliva, I. Hussein, M. Iqbal, J. Kimutai, T. Knowles, F. Meza, D. Nedjraoui, F. Tena, M. Türkeş, R. Vázquez, and M. Weltz. (2019). *Desertification. In: Climate Change and Land: an IPCC special report on climate change, desertification, land degradation, sustainable land management, food security, and greenhouse gas fluxes in terrestrial ecosystems*. Cambridge: Cambridge University Press.

[385] Mishra, S., I. Tewari, and S. Toosi. (2020). "Economic complexity and the globalization of services". *Structural Change and Economic Dynamics*. 53: 267–280.

[386] Mitchell, S. M. and C. B. Seaman. (2009). "A comparison of software cost, duration, and quality for waterfall vs. iterative and incremental development: A systematic review". In: *2009 3rd International Symposium on Empirical Software Engineering and Measurement*. IEEE. 511–515.

[387] Moeckel, C., K. Breivik, T. H. Nøst, A. Sankoh, K. C. Jones, and A. Sweetman. (2020). "Soil pollution at a major West African E-waste recycling site: Contamination pathways and implications for potential mitigation strategies". *Environment international*. 137: 105563.

[388] Mojzsis, S. J., G. Arrhenius, K. McKeegan, T. Harrison, A. Nutman, and C. Friend. (1996). "Evidence for life on Earth before 3,800 million years ago". *Nature*. 384(6604): 55–59.

[389] Moore, G. E. *et al.* (1965). "Cramming more components onto integrated circuits".

[390] Moore, K. J. and D. C. Lewis. (2000). "Multinational enterprise in ancient Phoenicia". *Business History*. 42(2): 17–42.

[391] Mounier, A. and M. M. Lahr. (2019). "Deciphering African late middle Pleistocene hominin diversity and the origin of our species". *Nature communications*. 10(1): 1–13.

[392] Müller, V. C. and N. Bostrom. (2016). "Future progress in artificial intelligence: A survey of expert opinion". *Fundamental issues of artificial intelligence*: 555–572.

[393] Müller-Pohl, S. (2010). "By Atlantic Telegraph". *Social Research*. 35(1): 9.

[394] Museum – International Print Museum. "Gutenberg Press replica". URL: https://www.printmuseum.org/gutenbergpress (accessed on 09/24/2022).

[395] N.N. "Episode 162 Beavers are Busy". URL: https://www.allcreaturespod.com/episodes/episode-162-beavers-are-busy/ (accessed on 06/05/2022).

[396] Näslund, D. (2008). "Lean, six sigma and lean sigma: fads or real process improvement methods?" *Business process management journal*.

[397] Nations – United Nations. (2015). *Paris Agreement*. United Nations.

[398] Nations – United Nations. (2019). *Ensure healthy lives and promote well-being for all at all ages*. Department, of Economics and Social Affairs, United Nations.

[399] Nations – United Nations. (2020). *United Nations Conference on Trade and Development (UNCTAD) Handbook of Statistics*. United Nations, Geneva.

[400] Naumann, J. D. and A. M. Jenkins. (1982). "Prototyping: the new paradigm for systems development". *Mis Quarterly*: 29–44.

[401] Nescio, N. (1864). *The Cotton Mill*. The Bolton Chronicle, Exeter University, UK. URL: https://cottonfaminepoetry.exeter.ac.uk/ (accessed on 10/16/2023).

[402] Nezafati, N. and T. Stöllner. (2017). "Economic geology, mining archaeological and archaeometric investigations at the Veshnaveh ancient copper mine, Central Iran". *Metalla*. 23(2): 67–90.

[403] Niinimäki, K., G. Peters, H. Dahlbo, P. Perry, T. Rissanen, and A. Gwilt. (2020). "The environmental price of fast fashion". *Nature Reviews Earth & Environment*. 1(4): 189–200.

[404] Nixon, G. (2021). *The world of bees*.

[405] Nolet, B. A. and F. Rosell. (1998). "Comeback of the beaver Castor fiber: an overview of old and new conservation problems". *Biological conservation*. 83(2): 165–173.

[406] Nonnenmacher, T. (2001). "History of the US Telegraph Industry". *EH. Net Encyclopedia, edited by Robert Whaples*.

[407] Nutman, A. P., V. C. Bennett, C. R. Friend, M. J. Van Kranendonk, and A. R. Chivas. (2016). "Rapid emergence of life shown by discovery of 3,700-million-year-old microbial structures". *Nature*. 537(7621): 535–538.

[408] Nye, D. E. (2013). *America's assembly line*. MIT Press.

[409] O'Driscoll, C. A. and J. C. Thompson. (2018). "The origins and early elaboration of projectile technology". *Evolutionary Anthropology: Issues, News, and Reviews*. 27(1): 30–45.

[410] O'Neill, B. C., E. Kriegler, K. L. Ebi, E. Kemp-Benedict, K. Riahi, D. S. Rothman, B. J. Van Ruijven, D. P. Van Vuuren, J. Birkmann, K. Kok, *et al.* (2017). "The roads ahead: Narratives for shared socioeconomic pathways describing world futures in the 21st century". *Global environmental change*. 42: 169–180.

[411] O'Neill, B. C., E. Kriegler, K. Riahi, K. L. Ebi, S. Hallegatte, T. R. Carter, R. Mathur, and D. P. Van Vuuren. (2014). "A new scenario framework for climate change research: the concept of shared socioeconomic pathways". *Climatic change*. 122: 387–400.

[412] O'Neill, D. W., A. L. Fanning, W. F. Lamb, and J. K. Steinberger. (2018). "A good life for all within planetary boundaries". *Nature sustainability*. 1(2): 88–95.

[413] O'rourke, K. H. and J. G. Williamson. (2002). "When did globalisation begin?" *European Review of Economic History*. 6(1): 23–50.

[414] Oates, J. (1993). "Trade and power in the fifth and fourth millennia BC: new evidence from northern Mesopotamia". *World Archaeology*. 24(3): 403–422.

[415] Obama, B. "Remarks by the President at U.N. Climate Change Summit". URL: https://obamawhitehouse.archives.gov/the-press-office/2014/09/23/remarks-president-un-climate-change-summit (accessed on 05/13/2022).

[416] Odling-Smee, F. J., K. N. Laland, and M. W. Feldman. (1996). "Niche construction". *The American Naturalist*. 147(4): 641–648.

[417] Oelze, N., T. Gruchmann, and M. Brandenburg. (2020). "Motivating factors for implementing apparel certification schemes—a sustainable supply chain management perspective". *Sustainability*. 12(12): 4823.

[418] Ohno, T. (2019). *Toyota production system: beyond large-scale production*. Productivity press.

[419] Olsen, I. T. (1998). "Sustainability of health care: a framework for analysis". *Health policy and planning*. 13(3): 287–295.

[420] Organization – International Labor Organization. "Decent Work". URL: https://www.ilo.org/global/topics/decent-work/lang--en/index.htm (accessed on 05/27/2023).

[421] Organization – World Health Organization. (2014). *Compendium of innovative health technologies for low-resource settings: assistive devices, eHealth solutions, medical devices*. World Health Organization.

[422] Organization – World Health Organization. (2017). "Environmentally sustainable health systems: a strategic document". *Tech. rep.* World Health Organization. Regional Office for Europe.

[423] Orlicky, O. J. (1975). "MRP, The new way of life in production and inventory management".

[424] Orlicky, J. A. (1973). "Net change material requirements planning". *IBM Systems Journal*. 12(1): 2–29.

[425] Ortiz-Ospina, E. and N. Lippolis. "Structural transformation – how did todays rich countriees become deindustrialized?". URL: https://ourworldindata.org/structural-transformation-and-deindustrialization-evidence-from-todays-rich-countries (accessed on 12/31/2022).

[426] Overton, M. (1996). *Agricultural revolution in England: the transformation of the agrarian economy 1500–1850*, No. 23. Cambridge University Press.

[427] Owusu, S., A. Szirmai, N. Foster-McGregor, *et al.* (2020). "The rise of the service sector in the global economy". *UNU-MERIT Working Paper Series*. 56.

[428] Oxford. "Dictionary". URL: https://www.oed.com (accessed on 06/14/2021).

[429] Page, R. E. and C. Y.-S. Peng. (2001). "Aging and development in social insects with emphasis on the honey bee, Apis mellifera L." *Experimental Gerontology*. 36(4): 695–711.

[430] Pamuk, S., J. L. van Zanden, S. Broadberry, and K. H. O'Rourke. (2010). "Standards of living". *The Cambridge economic history of modern Europe*: 217–234.

[431] Panter-Brick, C. (2002). "Sexual division of labor: energetic and evolutionary scenarios". *American Journal of Human Biology*. 14(5): 627–640.

[432] Parasuraman, A., V. A. Zeithaml, and L. Berry. (1988). "SERVQUAL: A multiple-item scale for measuring consumer perceptions of service quality". *1988*. 64(1): 12–40.

[433] Parliament, U. "UK Parliament Online". URL: https://www.parliament.uk (accessed on 08/20/2021).

[434] Parzinger, H. (2008). "The 'Silk Roads' concept reconsidered: about transfers, transportation and transcontinental interactions in prehistory". *The Silk Road*. 5(2): 7–15.

[435] Pastur, G. M., M. V. Lencinas, J. Escobar, P. Quiroga, L. Malmierca, and M. Lizarralde. (2006). "Understorey succession in Nothofagus forests in Tierra del Fuego (Argentina) affected by Castor canadensis". *Applied Vegetation Science*. 9(1): 143–154.

[436] Pätäri, S., A. Tuppura, A. Toppinen, and J. Korhonen. (2016). "Global sustainability megaforces in shaping the future of the European pulp and paper industry towards a bioeconomy". *Forest Policy and Economics*. 66: 38–46.

[437] Pathak, M., R. Slade, R. Pichs-Madruga, and D. Ürge-Vorsatz. (2022). "Technical Summary. In: Climate Change 2022: Mitigation of Climate

Change. Contribution of Working Group III to the Sixth Assessment Report of the Intergovernmental Panel on Climate Change".

[438] Paxton, J. (2012). "Mr. Taylor, Mr. Ford, and the Advent of High-Volume Mass Production: 1900–1912". *Economics & Business Journal*. 4(1): 74–90.

[439] Pearce, E., C. Stringer, and R. I. Dunbar. (2013). "New insights into differences in brain organization between Neanderthals and anatomically modern humans". *Proceedings of the Royal Society B: Biological Sciences*. 280(1758): 20130168.

[440] Peck, A. L. (1965). "Aristotle: Historia Animalium". *II (Cambridge MA, London)*.

[441] Pedroletti, D. and F. Ciabuschi. (2023). "Reshoring: A review and research agenda". *Journal of Business Research*. 164: 114005.

[442] Pegler, L., K. A. Siegmann, and S. Vellema. (2011). "Labour in globalized agricultural value chains". *Value chains, social inclusion and economic development: Contrasting theories and realities*: 102–120.

[443] Persson, L., B. M. Carney Almroth, C. D. Collins, S. Cornell, C. A. de Wit, M. L. Diamond, P. Fantke, M. Hassello?v, M. MacLeod, M. W. Ryberg, *et al.* (2022). "Outside the safe operating space of the planetary boundary for novel entities". *Environmental science & technology*. 56(3): 1510–1521.

[444] Persson, T. and G. Tabellini. (1991). "Is inequality harmful for growth? Theory and evidence".

[445] Petersen, K., C. Wohlin, and D. Baca. (2009). "The waterfall model in large-scale development". In: *Product-Focused Software Process Improvement: 10th International Conference, PROFES 2009, Oulu, Finland, June 15–17, 2009. Proceedings 10*. Springer. 386–400.

[446] Pfister, U. (2011). "Economic growth in Germany, 1500–1850". In: *conference on quantifying long run economic development, Venice*. 22–24.

[447] Pichler, P.-P., I. S. Jaccard, U. Weisz, and H. Weisz. (2019). "International comparison of health care carbon footprints". *Environmental research letters*. 14(6): 064004.

[448] Piercy, N. and N. Rich. (2015). "The relationship between lean operations and sustainable operations". *International Journal of Operations & Production Management*. 35(2): 282–315.

[449] Pinhasi, R., M. G. Thomas, M. Hofreiter, M. Currat, and J. Burger. (2012). "The genetic history of Europeans". *Trends in Genetics*. 28(10): 496–505.

[450] Pohjola, M. (2000). "Information technology and economic growth: A cross-country analysis".

[451] Pollard, S. (1980). "A new estimate of British coal production, 1750–1850". *The Economic History Review*. 33(2): 212–235.

[452] Pontzer, H. (2012). "Overview of Hominin Evolution". *Nature Education Knowledge.* 3(10).

[453] Poppendieck, M. and T. Poppendieck. (2003). *Lean software development: an agile toolkit.* Addison-Wesley.

[454] Porciello, J., M. Ivanina, M. Islam, S. Einarson, and H. Hirsh. (2020). "Accelerating evidence-informed decision-making for the Sustainable Development Goals using machine learning". *Nature Machine Intelligence.* 2(10): 559–565.

[455] Porter, M. E. (1996). "Operational effectiveness is not strategy". *Harvard Business Review.* 74(6): 61–78.

[456] Pörtner, H.-O., D. Roberts, M. Tignor, E. Poloczanska, K. Mintenbeck, A. Alegría, M. Craig, S. Langsdorf, S. Löschke, V. Möller, A. Okem, B. Rama, and (eds.) (2022). *Climate Change 2022: Impacts, Adaptation and Vulnerability. Contribution of Working Group II to the Sixth Assessment Report of the Intergovernmental Panel on Climate Change.* Press. Cambridge University Press, Cambridge, UK and New York, NY, USA.

[457] Potts, S. G., P. Neumann, B. Vaissière, and N. J. Vereecken. (2018). "Robotic bees for crop pollination: Why drones cannot replace biodiversity". *Science of the total environment.* 642: 665–667.

[458] Powell, D. (2013). "ERP systems in lean production: new insights from a review of lean and ERP literature". *International Journal of Operations & Production Management.*

[459] Prado, A., F. Requier, D. Crauser, Y. Le Conte, V. Bretagnolle, and C. Alaux. (2020). "Honeybee lifespan: the critical role of pre-foraging stage". *Royal Society open science.* 7(11): 200998.

[460] Price, C. "The Age of Scurvy". URL: www.sciencehistory.org (accessed on 10/09/2022).

[461] Programme, United Nations Environment Programme. (2023). "Tourism". URL: https://www.unep.org/explore-topics/resource-efficiency/what-we-do/responsible-industry/tourism (accessed on 08/12/2023).

[462] Pullin, A. S. (2002). *Conservation Biology.* Cambridge University Press. DOI: 10.1017/CBO9781139051927.

[463] Qin, L., M. Y. Malik, K. Latif, Z. Khan, A. W. Siddiqui, and S. Ali. (2021). "The salience of carbon leakage for climate action planning: Evidence from the next eleven countries". *Sustainable Production and Consumption.* 27: 1064–1076.

[464] Quandt, A., J. D. Salerno, J. C. Neff, T. D. Baird, J. E. Herrick, J. T. McCabe, E. Xu, and J. Hartter. (2020). "Mobile phone use is associated with higher smallholder agricultural productivity in Tanzania, East Africa". *PLoS One.* 15(8): e0237337.

[465] Rabassa, J., A. Coronato, G. Bujalesky, M. Salemme, C. Roig, A. Meglioli, C. Heusser, S. Gordillo, F. Roig, A. Borromei, *et al.* (2000). "Quaternary of Tierra del Fuego, southernmost South America: an updated review". *Quaternary International.* 68: 217–240.

[466] Raff, D. M. (1988). "Wage determination theory and the five-dollar day at Ford". *The Journal of Economic History.* 48(2): 387–399.

[467] Rajgopal, J. (2004). "Principles and applications of operations research". *Maynard's Industrial Engineering Handbook.–2004.–P*: 11–27.

[468] Rajlich, V. T. and K. H. Bennett. (2000). "A staged model for the software life cycle". *Computer.* 33(7): 66–71.

[469] Ramadan, B. S., I. Rachman, N. Ikhlas, S. B. Kurniawan, M. F. Miftahadi, and T. Matsumoto. (2022). "A comprehensive review of domestic-open waste burning: recent trends, methodology comparison, and factors assessment". *Journal of Material Cycles and Waste Management.* 24(5): 1633–1647.

[470] Ravindran, R. D., S. Gupta, A. Haripriya, T. Ravilla, and G.-B. B. Subburaman. (2021). "Seven-year trends in cataract surgery indications and quality of outcomes at Aravind Eye Hospitals, India". *Eye.* 35(7): 1895–1903.

[471] Rawlinson, G. (2019). *History of phoenicia.* Good Press.

[472] Raworth, K. (2012). *A safe and just space for humanity: can we live within the doughnut?* Oxfam.

[473] Raworth, K. (2017). "A Doughnut for the Anthropocene: humanity's compass in the 21st century". *The lancet planetary health.* 1(2): e48–e49.

[474] Reap, J., F. Roman, S. Duncan, and B. Bras. (2008). "A survey of unresolved problems in life cycle assessment: Part 2: impact assessment and interpretation". *The International Journal of Life Cycle Assessment.* 13: 374–388.

[475] Reichl, C. and M. Schatz. (2020). *World mining data.* Federal Ministry of Agriculture, Regions and Tourism, Vienna, Austria.

[476] Reimer, J. "Total share: 30 years of personal computer market share figures". URL: https://arstechnica.com/features/2005/12/total-share (accessed on 10/21/2021).

[477] Rejeb, A., K. Rejeb, S. J. Simske, and H. Treiblmaier. (2021). "Drones for supply chain management and logistics: a review and research agenda". *International Journal of Logistics Research and Applications*: 1–24.

[478] Riahi, K., D. P. Van Vuuren, E. Kriegler, J. Edmonds, B. C. O'neill, S. Fujimori, N. Bauer, K. Calvin, R. Dellink, O. Fricko, *et al.* (2017). "The Shared Socioeconomic Pathways and their energy, land use, and greenhouse gas emissions implications: An overview". *Global environmental change.* 42: 153–168.

[479] Riera, J. A., R. M. Lima, and O. M. Knio. (2023). "A review of hydrogen production and supply chain modeling and optimization". *International Journal of Hydrogen Energy*.

[480] Ritchie, H. and M. Roser. "Atmospheric concentrations". URL: https://ourworldindata.org/atmospheric-concentrations (accessed on 04/24/2023).

[481] Ritchie, H. and M. Roser. "Greenhouse Gas Emissions". URL: https://ourworldindata.org/co2-and-greenhouse-gas-emissions (accessed on 02/18/2023).

[482] Roberts, B. (2008). "Creating traditions and shaping technologies: understanding the earliest metal objects and metal production in Western Europe". *World Archaeology*. 40(3): 354–372.

[483] Roberts, B. W., C. P. Thornton, and V. C. Pigott. (2009). "Development of metallurgy in Eurasia". *Antiquity*. 83(322): 1012–1022.

[484] Robins, N. (2012). *The corporation that changed the world: how the East India Company shaped the modern multinational.* Pluto press.

[485] "Robots and their contribution to aged care". (2020). URL: https://magentahealthjapan.com/robots-their-contribution-to-aged-care/ (accessed on 09/25/2023).

[486] Rocha-Santos, T. A. (2018). "Editorial overview: micro and nano-plastics". *Current Opinion in Environmental Science & Health*. 1: 52–54.

[487] Rockström, J., W. Steffen, K. Noone, Å. Persson, F. S. Chapin, E. F. Lambin, T. M. Lenton, M. Scheffer, C. Folke, H. J. Schellnhuber, *et al.* (2009). "A safe operating space for humanity". *nature*. 461(7263): 472–475.

[488] Roebroeks, W. and P. Villa. (2011). "On the earliest evidence for habitual use of fire in Europe". *Proceedings of the National Academy of Sciences*. 108(13): 5209–5214. DOI: 10.1073/pnas.1018116108.

[489] Rogers, E. M., A. Singhal, and M. M. Quinlan. (2014). "Diffusion of innovations". In: *An integrated approach to communication theory and research*. Routledge. 432–448.

[490] Rose, D. C. and J. Chilvers. (2018). "Agriculture 4.0: Broadening responsible innovation in an era of smart farming". *Frontiers in Sustainable Food Systems*. 2: 87.

[491] Rost, S. (2017). "Water management in Mesopotamia from the sixth till the first millennium BC". *Wiley Interdisciplinary Reviews: Water*. 4(5): e1230.

[492] Roubik, D. W. (1992). *Ecology and natural history of tropical bees*. Cambridge University Press.

[493] Roux, V. and P. de Miroschedji. (2009). "Revisiting the History of the Potter's Wheel in the Southern Levant". *Levant*. 41(2): 155–173.

[494] Rowley-Conwy, P. (2011). "Westward Ho! The spread of agriculture from Central Europe to the Atlantic". *Current anthropology*. 52(S4): S431–S451.

[495] Rozhkova-Timina, I., V. Popkov, P. Mitchell, and S. Kirpotin. (2018). "Beavers as ecosystem engineers–a review of their positive and negative effects". In: *IOP Conference Series: Earth and Environmental Science*. Vol. 201. No. 1. IOP Publishing. 012015.

[496] Ruddiman, W. F. and E. C. Ellis. (2009). "Effect of per-capita land use changes on Holocene forest clearance and CO2 emissions". *Quaternary Science Reviews*. 28(27–28): 3011–3015.

[497] Ruedemann, R. and W. Schoonmaker. (1938). "Beaver-dams as geologic agents". *Science*. 88(2292): 523–525.

[498] Russell, S. (2019). *Human compatible: Artificial intelligence and the problem of control*. Penguin.

[499] Rybczynski, N. (2007). "Castorid phylogenetics: implications for the evolution of swimming and tree-exploitation in beavers". *Journal of Mammalian Evolution*. 14(1): 1–35.

[500] Sachs, J., C. Kroll, G. Lafortune, G. Fuller, and F. Woelm. (2022). *Sustainable development report 2022*. Cambridge University Press.

[501] Sackett, D. L., W. M. Rosenberg, J. M. Gray, R. B. Haynes, and W. S. Richardson. (1996). "Evidence based medicine: what it is and what it isn't".

[502] Şahoğlu, V. (2005). "The Anatolian trade network and the Izmir region during the Early Bronze Age". *Oxford Journal of Archaeology*. 24(4): 339–361.

[503] Samman, E., J. Roche, M. Binat Sarwar, and M. Evans. (2021). "Leave noone behind – 5 years into agenda 2030".

[504] Sandgathe, D. M. (2017). "Identifying and describing pattern and process in the evolution of hominin use of fire". *Current Anthropology*. 58(S16): S360–S370.

[505] Sassi, F. (2006). "Calculating QALYs, comparing QALY and DALY calculations". *Health policy and planning*. 21(5): 402–408.

[506] Sax, H., B. Allegranzi, M.-N. Chraïti, J. Boyce, E. Larson, and D. Pittet. (2009). "The World Health Organization hand hygiene observation method". *American journal of infection control*. 37(10): 827–834.

[507] Scheidel, W. (1997). "Quantifying the sources of slaves in the early Roman empire". *The Journal of Roman Studies*. 87: 156–169.

[508] Schettkat, R. and L. Yocarini. (2006). "The shift to services employment: A review of the literature". *Structural change and economic dynamics*. 17(2): 127–147.

[509] Schmandt-Besserat, D. (1982). "The emergence of recording". *American anthropologist*. 84(4): 871–878.

[510] Schmenner, R. (1993). *Production/Operations Management*. MacMillan Publishing Company.

[511] Schmenner, R. W. (2015). "The pursuit of productivity". *Production and Operations Management*. 24(2): 341–350.

[512] Schneider, F. (2012). "The shadow economy and work in the shadow: what do we (not) know?"

[513] Schneider, F. and D. H. Enste. (2013). *The shadow economy: An international survey*. Cambridge University Press.

[514] School, Oxford Martin School. "World GDP over the last two millenia". URL: https://www.ourworldindata.org (accessed on 07/09/2021).

[515] Schroeder, H. (1923). *History of electric light*. Vol. 76. No. 8–13. Smithsonian institution.

[516] Schumpeter, J. A. "Capitalism... is by nature or method of economic change and not only never is but never can be stationary."

[517] Schwab, K. (2016). *The Fourth Industrial Revolution*. World Economic Forum, Geneva, Switzerland.

[518] Scientists, Independent Group of Scientists appointed by the Secretary-General of the United Nations. (2019). *Global Sustainable Development Report 2019: The Future is Now – Science for Achieving Sustainable Development*. United Nations.

[519] Seeley, T. D. (2019). *The Lives of Bees*. Princeton University Press.

[520] Shah, A. A., F. Hasan, A. Hameed, and S. Ahmed. (2008). "Biological degradation of plastics: a comprehensive review". *Biotechnology advances*. 26(3): 246–265.

[521] Shen, J., L. Yuan, J. Zhang, H. Li, Z. Bai, X. Chen, W. Zhang, and F. Zhang. (2011). "Phosphorus dynamics: from soil to plant". *Plant physiology*. 156(3): 997–1005.

[522] Shukla, P., J. Skea, A. Reisinger, R. Slade, R. Fradera, M. Pathak, A. Al Khourdajie, M. Belkacemi, R. Van Diemen, A. Hasija, G. Lisboa, S. Luz, J. Malley, D. McCollum, S. Some, P. Vyas, and (eds.) (2023). "Synthesis report of the IPCC sixth assessment report (AR6)–Summary for Policymakers". In: *Climate Change 2022: Mitigation of Climate Change. Contribution of Working Group III to the Sixth Assessment Report of the Intergovernmental Panel on Climate Change*. Cambridge University Press, Cambridge, UK and New York, NY, USA.

[523] Shukla, P., J. Skea, R. Slade, A. Al Khourdajie, R. van Diemen, D. McCollum, M. Pathak, S. Some, P. Vyas, R. Fradera, M. Belkacemi, A. Hasija, G. Lisboa, S. Luz, J. Malley, and (eds.) (2022). *Climate Change 2022: Mitigation of Climate Change. Working Group III contribution to the Sixth Assessment Report of the Intergovernmental Panel on Climate Change*. Cambridge University Press, Cambridge, UK and New York, NY, USA.

[524] Silver, H. (2015). "The contexts of social inclusion". *Available at SSRN 2641272.*

[525] Smil, V. (2000). "Energy in the twentieth century: resources, conversions, costs, uses, and consequences". *Annual Review of Energy and the Environment.* 25(1): 21–51.

[526] Smith, C. W. and J. T. Cothren. (1999). *Cotton: origin, history, technology, and production.* Vol. 4. John Wiley & Sons.

[527] Smith, F. A., R. E. Elliott Smith, S. K. Lyons, and J. L. Payne. (2018). "Body size downgrading of mammals over the late Quaternary". *Science.* 360(6386): 310–313.

[528] Smith, T. (2020). "Merthyr Tydfil–UK's iron metropolis". *Steel Times International.* 44(4): 48–48.

[529] Soeder, D. J. (2020). *Fracking and the Environment: A Scientific Assessment of the Environmental Risks from Hydraulic Fracturing and Fossil Fuels.* Springer.

[530] Solow, R. (1987). "We'd better watch out". *New York Times Book Review.* 36.

[531] Sott, M. K., L. B. Furstenau, L. M. Kipper, F. D. Giraldo, J. R. Lopez-Robles, M. J. Cobo, A. Zahid, Q. H. Abbasi, and M. A. Imran. (2020). "Precision techniques and agriculture 4.0 technologies to promote sustainability in the coffee sector: state of the art, challenges and future trends". *IEEE Access.* 8: 149854–149867.

[532] Soundararajan, V., L. J. Spence, and C. Rees. (2018). "Small business and social irresponsibility in developing countries: Working conditions and "evasion" institutional work". *Business & Society.* 57(7): 1301–1336.

[533] Sousa Jabbour, A. B. L. de, C. J. C. Jabbour, M. Godinho Filho, and D. Roubaud. (2018). "Industry 4.0 and the circular economy: a proposed research agenda and original roadmap for sustainable operations". *Annals of Operations Research.* 270(1): 273–286.

[534] Spear, B. (2008). "James Watt: The steam engine and the commercialization of patents". *World patent information.* 30(1): 53–58.

[535] Spitzer, J. (2017). "Emergence of life on earth: a physicochemical jigsaw puzzle". *Journal of molecular evolution.* 84(1): 1–7.

[536] Spottiswoode, C. N., K. S. Begg, and C. M. Begg. (2016). "Reciprocal signaling in honeyguide-human mutualism". *Science.* 353(6297): 387–389.

[537] Springmann, M., M. Clark, D. Mason-D'Croz, K. Wiebe, B. L. Bodirsky, L. Lassaletta, W. De Vries, S. J. Vermeulen, M. Herrero, K. M. Carlson, *et al.* (2018). "Options for keeping the food system within environmental limits". *Nature.* 562(7728): 519–525.

[538] Starna, W. A. and J. A. Brandão. (2004). "From the Mohawk–Mahican War to the Beaver Wars: Questioning the Pattern". *Ethnohistory.* 51(4): 725–750.

[539] Statista. "Global Mobile Messaging Traffic". URL: https://www.statista.com/statistics/483255/numberofmobilemessaginguserworldwide (accessed on 12/29/2022).

[540] Statista. "Global MobileEmail Traffic". URL: https://www.statista.com/statistics/456500/dailynumberofemailsworldwide (accessed on 12/29/2022).

[541] Statista. "Global Shipments of Personal Computers". URL: https://www.statista.com/statistics/273495/global-shipments-of-personal-computers-since-2006/ (accessed on 10/21/2021).

[542] Statistics – Central Bureau for Statistics, The Netherlands. "Zorginstellingen: financien en personeel". URL: https://www.cbs.nl/nl-nl/cijfers/detail/83626NED (accessed on 02/12/2023).

[543] Steffen, W., K. Richardson, J. Rockström, S. E. Cornell, I. Fetzer, E. M. Bennett, R. Biggs, S. R. Carpenter, W. De Vries, C. A. De Wit, *et al.* (2015). "Planetary boundaries: Guiding human development on a changing planet". *Science*. 347(6223): 1259855.

[544] Stern, D. I. and A. Kander. (2012). "The role of energy in the industrial revolution and modern economic growth". *The Energy Journal*. 33(3).

[545] Stewart, Q. T. and J. C. Dixon. (2010). "Is it race, immigrant status, or both? An analysis of wage disparities among men in the United States". *International Migration Review*. 44(1): 173–201.

[546] "Structural transformation – how did todays rich countries become deindustrialized?" URL: https://ourworldindata.org/structural-transformation-and-deindustrialization-evidence-from-todays-rich-countries (accessed on 12/31/2022).

[547] Styles, J. (2020). "The Rise and Fall of the Spinning Jenny: Domestic Mechanisation in Eighteenth-Century Cotton Spinning". *Textile History*. 51(2): 195–236.

[548] Talberth, J. and M. Weisdorf. (2017). "Genuine progress indicator 2.0: pilot accounts for the US, Maryland, and City of Baltimore 2012–2014". *Ecological Economics*. 142: 1–11.

[549] Tansel, B. (2017). "From electronic consumer products to e-wastes: Global outlook, waste quantities, recycling challenges". *Environment international*. 98: 35–45.

[550] Tattersall, I. (2009). "Human origins: out of Africa". *Proceedings of the National Academy of Sciences*. 106(38): 16018–16021.

[551] Taylor, F. W. (2004). *Scientific management*. Routledge.

[552] Taylor, T. (2001). "Believing the ancients: quantitative and qualitative dimensions of slavery and the slave trade in later prehistoric Eurasia". *World Archaeology*. 33(1): 27–43.

[553] Team, Core Writing Team, and Pachauri, R.K. and Meyer, L.A. and (eds.) (2014). *Climate Change 2014: Synthesis Report. Contribution of Working Groups I, II and III to the Fifth Assessment Report of the Intergovernmental Panel on Climate Change.* IPCC, Geneva, Switzerland.

[554] Terhorst, A. and J. A. Dowling. (2022). "Terrestrial analogue research to support human performance on Mars: A review and bibliographic analysis". *Space: Science & Technology.* 2022.

[555] Thomas, D. P. (1997). "Sailors, scurvy and science". *Journal of the Royal Society of Medicine.* 90(1): 50–54.

[556] Thompson, J. C., D. K. Wright, and S. J. Ivory. (2021). "The emergence and intensification of early hunter-gatherer niche construction". *Evolutionary Anthropology: Issues, News, and Reviews.* 30(1): 17–27.

[557] Thornton, R. (1984). "Cherokee population losses during the Trail of Tears: A new perspective and a new estimate". *Ethnohistory*: 289–300.

[558] Toop, T. A., S. Ward, T. Oldfield, M. Hull, M. E. Kirby, and M. K. Theodorou. (2017). "AgroCycle–developing a circular economy in agriculture". *Energy Procedia.* 123: 76–80.

[559] "Transforming Dreams into Sounds". (2023). URL: https://solarear.com.br/.

[560] Trauth, M. H., M. A. Maslin, A. L. Deino, A. Junginger, M. Lesoloyia, E. O. Odada, D. O. Olago, L. A. Olaka, M. R. Strecker, and R. Tiedemann. (2010). "Human evolution in a variable environment: the amplifier lakes of Eastern Africa". *Quaternary Science Reviews.* 29(23–24), 2981–2988.

[561] Council, World Travel and Tourism Council. (2023). "Economic Impact Research". URL: https://wttc.org/research/economic-impact (accessed on 08/12/2023).

[562] Triplett, J. E. and B. P. Bosworth. "Productivity in the services sector". In: *presentation at the American Economic Association meetings, January.* Vol. 9. 301.

[563] Tripp, S. D. and B. Bichelmeyer. (1990). "Rapid prototyping: An alternative instructional design strategy". *Educational technology research and development*: 31–44.

[564] Turing, A. M. (1950). "Computing machinery and intelligence". *Mind.* 59: 433–460.

[565] Twomey, T. (2013). "The cognitive implications of controlled fire use by early humans". *Cambridge Archaeological Journal.* 23(1): 113–128.

[566] UNDO, United Nations Development Organization. (1996). *Human Development Report 1996.* Oxford University Press.

[567] United States. Department of Commerce, National Production Authority. (1953). *Historical Reports on Defense Production, Report 52, Communication Equipment.* US Government Printing Office.

[568] Ur, J. A., P. Karsgaard, and J. Oates. (2007). "Early urban development in the Near East". *Science.* 317(5842): 1188–1188.

[569] Valbuena-Carabaña, M., U. L. de Heredia, P. Fuentes-Utrilla, I. González-Doncel, and L. Gil. (2010). "Historical and recent changes in the Spanish forests: a socio-economic process". *Review of palaeobotany and palynology.* 162(3): 492–506.

[570] Van de Klundert, J. J. (2013). "Quality and Value of Health Services". In: *Service Management in Health and Wellness Services.* Ed. by K. J. Kendall Hunt Publishing.

[571] Van den Bergh, J. C. (2009). "The GDP paradox". *Journal of Economic Psychology.* 30(2): 117–135.

[572] Van der Kooij, B. (2015). "The Invention of the Electro-motive Engine".

[573] Van Engeland, J., J. Beliën, L. De Boeck, and S. De Jaeger. (2020). "Literature review: Strategic network optimization models in waste reverse supply chains". *Omega.* 91: 102012.

[574] Van Maanen, J., P. Frost, P. Moore, C. Lundberg, M. Louis, and J. Martin. (1991). *The smile factory.*

[575] Van Vuuren, D., E. Stehfest, D. Gernaat, H.-S. de Boer, V. Daioglou, J. Doelman, O. Edelenbosch, M. Harmsen, W.-J. van Zeist, M. van den Berg, *et al.* (2021). "The 2021 SSP scenarios of the IMAGE 3.2 model".

[576] Van Wassenhove, L. N. (2019). "Sustainable innovation: Pushing the boundaries of traditional operations management". *Production and Operations Management.* 28(12): 2930–2945.

[577] Van Zanden, J. L., J. Baten, P. Foldvari, and B. Van Leeuwen. (2014). "The changing shape of global inequality 1820–2000; exploring a new dataset". *Review of income and wealth.* 60(2): 279–297.

[578] Vanek, J. n., M. A. Chen, F. Carré, J. Heintz, and R. Hussmanns. (2014). "Statistics on the informal economy: Definitions, regional estimates and challenges". *Women in Informal Employment: Globalizing and Organizing (WIEGO) Working Paper (Statistics).* 2(1): 47–59.

[579] Varekamp, J. C. (2006). "The historic fur trade and climate change". *Eos, Transactions American Geophysical Union.* 87(52): 593–597.

[580] Vasan, C. S., S. Gupta, M. Shekhar, K. Nagu, L. Balakrishnan, R. D. Ravindran, T. Ravilla, and G.-B. B. Subburaman. (2023). "Accuracy of an artificial intelligence-based mobile application for detecting cataracts: Results from a field study". *Indian Journal of Ophthalmology.* 71(8): 2984–2989.

[581] Veal, A. J. (2004). "A brief history of work and its relationship to leisure". In: *Work and leisure*. Routledge. 15–33.

[582] Veblen, T. and C. W. Mills. (2017). *The theory of the leisure class*. Routledge.

[583] Velis, C. A. and E. Cook. (2021). "Mismanagement of plastic waste through open burning with emphasis on the global south: a systematic review of risks to occupational and public health". *Environmental Science & Technology*. 55(11): 7186–7207.

[584] Verderame, L. (2018). "Slavery in third-millennium Mesopotamia: an overview of sources and studies". *Journal of global slavery*. 3(1–2): 13–40.

[585] Vicol, M., J. Neilson, D. F. S. Hartatri, and P. Cooper. (2018). "Upgrading for whom? Relationship coffee, value chain interventions and rural development in Indonesia". *World Development*. 110: 26–37.

[586] Vigne, J.-D., F. Briois, A. Zazzo, G. Willcox, T. Cucchi, S. Thiébault, I. Carrère, Y. Franel, R. Touquet, C. Martin, *et al.* (2012). "First wave of cultivators spread to Cyprus at least 10,600 y ago". *Proceedings of the National Academy of Sciences*. 109(22): 8445–8449.

[587] Villarrubia-Gómez, P., S. E. Cornell, and J. Fabres. (2018). "Marine plastic pollution as a planetary boundary threat–The drifting piece in the sustainability puzzle". *Marine policy*. 96: 213–220.

[588] Vinuesa, R., H. Azizpour, I. Leite, M. Balaam, V. Dignum, S. Domisch, A. Felländer, S. D. Langhans, M. Tegmark, and F. Fuso Nerini. (2020). "The role of artificial intelligence in achieving the Sustainable Development Goals". *Nature communications*. 11(1): 233.

[589] Von Frisch, K. (2013). *Tanzsprache und orientierung der bienen*. Springer-Verlag.

[590] Waldman, E. and B. J. McEaddy. (1974). "Where women work—an analysis by industry and occupation". *Monthly Labor Review*: 3–13.

[591] Walker, H., S. Seuring, J. Sarkis, and R. Klassen. (2014). "Sustainable operations management: recent trends and future directions". *International Journal of Operations & Production Management*.

[592] Walker, M. (2017). *Why we sleep: the new science of sleep and dreams*. Penguin UK.

[593] Wang, F., J. D. Harindintwali, Z. Yuan, M. Wang, F. Wang, S. Li, Z. Yin, L. Huang, Y. Fu, L. Li, *et al.* (2021). "Technologies and perspectives for achieving carbon neutrality". *The Innovation*. 2(4): 100180.

[594] Watson, A. M. (1974). "The Arab agricultural revolution and its diffusion, 700–1100". *The Journal of Economic History*. 34(1): 8–35.

[595] Watson, E. C. (1945). "Richard Trevithick and the First Railway Locomotive". *Engineering and Science*. 8(9): 12–13.

[596] Watt, E., M. Picard, B. Maldonado, M. A. Abdelwahab, D. F. Mielewski, L. T. Drzal, M. Misra, and A. K. Mohanty. (2021). "Ocean plastics: environmental implications and potential routes for mitigation–a perspective". *RSC advances*. 11(35): 21447–21462.

[597] Weinberg, S. (1977). *The first three minutes A modern view of the origin of the universe*. Andre Deutsch, United Kingdom.

[598] Wendel, J. F. and R. C. Cronn. (2003). "Polyploidy and the evolutionary history of cotton". *Advances in agronomy*. 78: 139.

[599] Weninger, B., E. Alram-Stern, E. Bauer, L. Clare, U. Danzeglocke, O. Jöris, C. Kubatzki, G. Rollefson, H. Todorova, and T. van Andel. (2006). "Climate forcing due to the 8200 cal yr BP event observed at Early Neolithic sites in the eastern Mediterranean". *Quaternary Research*. 66(3): 401–420.

[600] Wenseleers, T. and F. L. Ratnieks. (2006). "Comparative analysis of worker reproduction and policing in eusocial Hymenoptera supports relatedness theory". *The American Naturalist*. 168(6): E163–E179.

[601] Whitehead, M. (1991). "The concepts and principles of equity and health". *Health promotion international*. 6(3): 217–228.

[602] Wichroski, M. (1994). "The secretary: Invisible labor in the workworld of women". *Human Organization*. 53(1): 33–41.

[603] Wilkinson, A., M. Hill, and P. Gollan. (2001). "The sustainability debate". *International Journal of Operations & Production Management*. 21(12): 1492–1502.

[604] Williams, C. C. and F. Schneider. (2016). *Measuring the Global Shadow Economy: the prevalence of informal work and labour*. Edward Elgar Publishing.

[605] Williams, K., C. Haslam, and J. Williams. (1992). "Ford versus Fordism". The Beginning of Mass Production?" *Work, Employment and Society*. 6(4): 517–555.

[606] Williams, K., C. Haslam, J. Williams, A. Adcroft, and S. Johal. (1993). "The myth of the line: Ford's production of the Model T at Highland Park, 1909–16". *Business History*. 35(3): 66–87.

[607] Wills Jr, J. E. (1993). "Maritime Asia, 1500-1800: The Interactive Emergence of European Domination". *The American Historical Review*. 98(1): 83–105.

[608] Wilson, J. M. (2011). "Reconsidering Ford's Highland Park Assembly Line: New Data Versus Old Ideas". In: *Academy of Management Proceedings*, Vol. 2011. No. 1. Academy of Management Briarcliff Manor, NY 10510. 1–6.

[609] Wilson, J. M. (2016). "The origin of material requirements planning in Frederick W. Taylor's planning office". *International Journal of Production Research*. 54(5): 1535–1553.

[610] Winston, B. (2002). *Media, technology and society: A history: From the telegraph to the Internet.* Routledge.

[611] Winston, M. L. (1991). *The biology of the honey bee.* Harvard University Press.

[612] Winston, M. L. and K. N. Slessor. (1998). "Honey bee primer pheromones and colony organization: gaps in our knowledge". *Apidologie.* 29(1–2): 81–95.

[613] Wirtz, J., W. Kunz, and S. Paluch. (2021). "The service revolution, intelligent automation and service robots". *European Business Review.* 29(5): 909.

[614] Women's Bureau, Records of the Women's Bureau, National Archives. "Switch Board Operators". URL: http://smithsonian.cowanrailroadmuseum.org/images.html (accessed on 02/01/2023).

[615] Wood, B. M., H. Pontzer, D. A. Raichlen, and F. W. Marlowe. (2014). "Mutualism and manipulation in Hadza–honeyguide interactions". *Evolution and Human Behavior.* 35(6): 540–546.

[616] Wood, L. C., C. Wang, H. Abdul-Rahman, and N. S. J. Abdul-Nasir. (2016). "Green hospital design: integrating quality function deployment and end-user demands". *Journal of Cleaner Production.* 112: 903–913.

[617] Wood, T. J., D. Michez, R. J. Paxton, M. Drossart, P. Neumann, M. Gerard, M. Vanderplanck, A. Barraud, B. Martinet, N. Leclercq, *et al.* (2020). "Managed honey bees as a radar for wild bee decline?" *Apidologie.* 51: 1100–1116.

[618] Woodbury, R. S. (1960). "The legend of Eli Whitney and interchangeable parts". *Technology and Culture.* 1(3): 235–253.

[619] Woods, C. and C. E. Woods. (2010). *Visible language: inventions of writing in the ancient Middle East and beyond; [in conjunction with the Exhibition Visible Language: Inventions of Writing in the Ancient Middle East and Beyond].* Oriental Institute of the University of Chicago.

[620] Woudhuizen, F. C. (2018). "Towards a reconstruction of Tin-trade routes in mediterranean protohistory". *Praehistorische Zeitschrift.* 92(2): 342–353.

[621] Wrangham, R. (2017). "Control of Fire in the Paleolithic: Evaluating the Cooking Hypothesis". *Current Anthropology.* 58(S16): S303–S313.

[622] Wrangham, R. W., J. H. Jones, G. Laden, D. Pilbeam, and N. Conklin-Brittain. (1999). "The raw and the stolen: cooking and the ecology of human origins". *Current anthropology.* 40(5): 567–594.

[623] WRI, World Resource Institute. "CAIT 2.0, Climate Analysis Indicators Tool: WRI's Climate Data Explorer". URL: http://cait2.wri.org (accessed on 09/30/2021).

[624] Wright, J. P., C. G. Jones, and A. S. Flecker. (2002). "An ecosystem engineer, the beaver, increases species richness at the landscape scale". *Oecologia.* 132(1): 96–101.

[625] Wrigley, C. (2011). "Out and about in Cromford Mill, Lea Mills and the Lumsdale Valley". *Historian* (111): 26.

[626] Wrigley, E. A. (2013). "Energy and the English industrial revolution". *Philosophical Transactions of the Royal Society A: Mathematical, Physical and Engineering Sciences*. 371(1986): 20110568.

[627] Xu, X., Y. Lu, B. Vogel-Heuser, and L. Wang. (2021). "Industry 4.0 and Industry 5.0—Inception, conception and perception". *Journal of Manufacturing Systems*. 61: 530–535.

[628] Xue, L., G. Liu, J. Parfitt, X. Liu, E. Van Herpen, Å. Stenmarck, C. O'Connor, K. O?stergren, and S. Cheng. (2017). "Missing food, missing data? A critical review of global food losses and food waste data". *Environmental science & technology*. 51(12): 6618–6633.

[629] Yafa, S. (2006). *Cotton: The biography of a revolutionary fiber*. Penguin.

[630] Yao, Z., Y. Lum, A. Johnston, L. M. Mejia-Mendoza, X. Zhou, Y. Wen, A. Aspuru-Guzik, E. H. Sargent, and Z. W. Seh. (2023). "Machine learning for a sustainable energy future". *Nature Reviews Materials*. 8(3): 202–215.

[631] Yearbook, China Statistical Yearbook. (2019). "National Bureau of statistics of China". *China Statistical Yearbook*.

[632] Yu, P., Z. Cai, and Y. Sun. (2021). "Does the emissions trading system in developing countries accelerate carbon leakage through OFDI? Evidence from China". *Energy Economics*. 101: 105397.

[633] Zamora, J. C. P. (2014). "The "Avoid-Shift-Improve" Model: A Powerful Planning Tool for Transportation Schemes with Low GHG Emissions". *MIPALCON 2014*: 141.

[634] Zeder, M. A. (2011). "The origins of agriculture in the Near East". *Current Anthropology*. 52(S4): S221–S235.

[635] Zeder, M. A. (2012). "The domestication of animals". *Journal of anthropological research*. 68(2): 161–190.

[636] Zimm, C., F. Sperling, and S. Busch. (2018). "Identifying sustainability and knowledge gaps in socio-economic pathways vis-à-vis the Sustainable Development Goals". *Economies*. 6(2): 20.

Index